E.H Ruffner

The Practice of the Improvement of the Non-Tidal Rivers of the

United States

E.H Ruffner

The Practice of the Improvement of the Non-Tidal Rivers of the United States

ISBN/EAN: 9783337233129

Printed in Europe, USA, Canada, Australia, Japan

Cover: Foto ©berggeist007 / pixelio.de

More available books at **www.hansebooks.com**

THE PRACTICE

OF THE

IMPROVEMENT

OF THE

NON-TIDAL RIVERS

OF THE UNITED STATES,

WITH AN EXAMINATION OF THE RESULTS THEREOF.

BY

CAPT. E. H. RUFFNER,

CORPS OF ENGINEERS, U. S. A.

NEW YORK:

JOHN WILEY & SONS.

1886.

TABLE OF CONTENTS.

ERRATA.

Page 16, line 21, omit "slight."

Page 28, 4th line from bottom, for "July 7th, 1887," read "July 5th, 1884."

Page 29, line 26, for "cutting way its," read "cutting its way."

Page 68 and 69, read "Yallabusha" wherever "Tallabusha," and in the second table on page 69, read "Tallahatchie" for "Yallahatchie."

Page 72, line 15, for "1884" read "1874."

Page 113, line 13 from the bottom, for "formed" read "found."

Page 124, line 4 from the bottom, should be a foot note.

Page 133, line 9 from the bottom, for "river will" read "river it will."

Page 135, line 11, for "coucentration" read "concentration."

Page 150, line 7, omit word "only."

Page 180, line 16, after "action" make a semi-colon.

Page 180, line 18, omit first two commas.

CHAPTER I.

THE present condition of the system of improvements of the rivers and harbors of the United States is somewhat anomalous. The practice of both political parties, since the war, has been one of liberal appropriations for that purpose, and beginning with 1866, general bills have passed, granting sums of more or less magnitude, every year, excepting three, 1877, 1883 and 1885; and in addition to these many specific bills for particular purposes were passed when emergencies called for prompt action. With the growth of the country and its expenses, and particularly in times of large revenue, the tendency has been to increase the number of objects for which money has been given, and also the amounts allotted to many specific and well known rivers, and harbors. But it has happened of late years that the opposition, which always existed, has increased in strength, *pari passu*, with the magnitude of the bills; and whether the components of this opposition base their action upon principle, jealousy, or opposing interests, they are now united in active and effective shape.

That the practice followed by both political parties, when in possession of one or both houses of Congress, during eighteen years past, of thus applying amounts varying from one and a half, to nineteen millions of dollars, should be suddenly dropped, because an active and intelligent opposition has shown, by argument, this practice to be wrong in principle, and inexpedient; can hardly be the fact.

That an organized and persistent minority can defeat legislation, to which it is opposed, has been repeatedly shown; especially during the closing hours of a Congress;—that this course would be pursued if it could be shown that such legislation were for the best interests of the country, is in the highest degree improbable.

That men from every section of the country should devote months of patient work in the preparation of these bills, and the necessary study of the subject; and that successive Congresses of different political parties, following the same course, should each and all of them wilfully conspire to waste the public money, or worse yet, misapply it; is logically and humanly impossible, and yet this assertion is shamelessly repeated year after year. But that errors may be made in the conception and execution of these plans, and the making of the bills, is but an instance of human fallibility.

That a business which legitimately calls for an expenditure of fourteen or nineteen millions of dollars in one year, can be safely left the next year without any resources, is a question of importance.

That there are great differences between the necessities for such public works, and great differences in the possible ways of best relieving these necessities, are also questions which should be considered.

It has seemed to me that the present is somewhat in the nature of a crisis in the general system of the improvement of rivers and harbors; and that many of the general propositions here presented, can well be discussed at this time.

As an officer of the Engineer Corps, acting in my official capacity, the custom of my profession forbids me to discuss all of these questions from the stand point of Congress, and I am further limited from commenting upon the engineering practice, or plans, authorized by official authority; or upon such matters as have been decided by the action of military superiors; yet as my position necessarily makes me familiar with many points, and particularly with the history of the progress of the works of improvement carried on under the Corps of Engineers, and the literature thereof; and as members of Congress, and the public at large, may not have either the time or inclination to study these; there are many points upon which a concise statement of facts must lead to conclusions, otherwise difficult to detect, and it is my intention to prepare and present such, in this work. Again, as a member of the Corps of Engineers, I have somewhat felt it my duty to present some of those questions from a stand point which could not be taken in official reports. Opposition to the river and harbor bill, and zeal in defence of it, have both resulted in attacks in reference to those in charge of these works, which can best be answered by clear statements of facts.

One of the difficulties in the preparation of the river and harbor bill is the amount of study needed to master the facts. The committee, appointed just before the holidays, is ready for work the first week in January, hardly before. By examining the regular bills in the long sessions of Congress, it is seen that the earliest dates of passage were June 10th, 1872, and June 15th, 1880; and in general this bill dates upon the last day of the session, or not more than two days before it. The careful preparation of the bill, must depend upon the accurate presentation of the facts upon which to base it. The Reports of the Chief of Engineers have of course increased in size with the number of items in the bill, and voluminous reports upon surveys of projected works are generally more bulky than reports of construction. During the years 1866 to 1873, one volume each year contained the report of the Chief of Engineers, varying from 630 to 1255 pages of contents. Steadily increasing since that time, the report for 1884 was made in four volumes, and covered 2903 pages. Besides the reports for the year, conscientious men will examine many manuscripts and maps, and take verbal testimony upon many points.

Although each annual report attempts to present the main facts in the history of each item of work, yet reference must be had to past volumes, which from 1867 to 1884, both included, embrace more than 32,000 pages. So long as the question of improvement remained one of expediency only, in the hands of the committee; the work of allotting the money, which could be granted, was simply one of places and amounts. But in the past, occasionally, and lately, often; questions of engineering methods and plans have been selected after examination by the committees, and debated in detail in Congress, thus involving much additional time and labor. Moreover, all these questions varied from year to year, and the precedent of one session did not necessarily solve the problem for the next. The labor by the committee could not of course in general be available in the house, except through its result; and when the bill came up on its passage, many conclusions of the committee were taken upon faith—or helped to injure the bill. The bill coming before the Senate late in the session, necessarily received less careful scrutiny than it would have done, had time permitted; and upon two occasions this was the real, if not the ostensible cause of the failure of its passage. Viewing the entire situation, and studying the history of the whole, the real surprise is, not that the bill has failed at times, but that it has passed as often as it has, and as carefully digested; solving upon many occasions knotty points of expediency and plans, in a successful manner. I will venture, later on, to make some suggestions as to the way in which the necessary information coming before the committees, can be condensed, and their work facilitated; and it is also a part of the intent of this essay to point out certain lines of facts, which, experience shows, indicate what limitations bound our system of improvements. Certain facts, fairly presented, will lead to inevitable conclusions.

There can be little doubt that had time permitted, the river and harbor bill of the last session, would not have failed. The close attention given by the committee to the bill, resulted in their thorough understanding of the subject, and adoption of their own conclusions; and the debates in the House show, that whatever the opinions, or votes of individual members might be, there was no reason they could not have been based upon a fairly clear idea of the provisions and intentions of the bill as outlined in the speeches. There is also little doubt that had time allowed, the bill would have been amended, item by item, in accordance with the wishes of the House, and then have passed;—and been a fair representation of the average, or compromise of sentiment, and wishes of the members; besides being in general based upon the reports of the engineers; afterwards acted upon by the committee, but corrected by the House itself.

The opposition to River and Harbor Bills in general, has been based upon two broad assertions. The first is made and repeated not once, but constantly by some, that it is a " grab," a "steal," "pork," and other less offensive terms intended to convey the impression that they are contrary to common principles of honesty, and designed in a spirit of corruption. It seems incredible that such a statement could be made. During the eighteen years past, there is not only no instance of a member of Congress being interested directly in a specific fraud in connection with these improvements, but no assertion has ever been made that these appropriations were ever used indirectly to the advantage of any congressman. Moreover it may be noted, that these unfounded assertions are only made at the time it is desired to defeat the bill in Congress. It would seem a waste of time to make this denial; but in the face of these annual assertions it will do no harm to say here, that a more baseless charge never was made. That members of Congress should bring to the notice of the committees the needs of their districts is as eminently proper in this matter, as in any other, in which they act as representatives.

Surveys for projected works, ordered by law; and heretofore made without an expression as to their value; were by the act of 1882, limited to those, where after a preliminary examination made by the engineer in charge, he reports "such river or harbor as worthy of improvement, and that the work is a public necessity;" when the survey is made, the act further directs, that "the commercial importance, present and prospective, of the improvement contemplated" shall be reported as fully as possible. The act of 1884 extends this principle, by stating that in the annual report of the Secretary of War as to the expenditures of the sums, "he shall report the effect of such work, together with such recommendations as he may deem proper to lay before Congress." It is evident that if this system of exact reports as to commercial importance, public necessity, and effect of works, together with suitable recommendations, be made by engineers and the Secretary of War, undue local influence must be eliminated; and the system may well be extended, with advantage, to embrace all works coming within the bill, and thus develop the full views of the Engineer Department, and the Administration.

A bill based upon such reports may be inexpedient, but it cannot be a "job." At this point attention may be called to the fact that if the congressional record be examined, it will be found that there is always a succession of memorials, and petitions from various commercial and legislative bodies, and from individuals, asking for the action of Congress on specific objects of river and harbor improvements; but no remonstrances will be found there either against the system as a whole, or asserting it to be a perversion of the public money, or that it is wasteful or extravagant.

The second ground of opposition is the most serious and practical. For the reason assigned, that the interests of the country did not justify or call for an appropriation of such magnitude, one executive suspended the expenditure of a portion of one river and harbor bill; and a second vetoed another for the same reason, and because in his judgment many of the items of the bill were not for objects of national importance, and did not justify the expenditures proposed. This ground has been held by many, and strongly supported by a portion of the press. To me it has seemed opportune at this time to examine somewhat carefully and in detail, this portion of the question. Believing that the whole river and harbor question is now simply a business matter, and depends entirely upon the amount of money which can be so invested, and the best way to invest it; I propose to review in detail, the progress and results, of the improvements of the non-tidal rivers of the United States, with a view of deducing principles for future application. To do this, it will be necessary to give the character of rivers and the means of improvement adopted; and it will be especially necessary to consider if the results commercially justified the expenditure. It will also be required to touch upon certain engineering points, which can be discussed under the limitations caused by my official position.

Some general remarks will be made upon the harbors, tidal rivers, and other cognate matters; but if it be recollected that the objectionable features of recent bills have been asserted as connected with the rivers, it will be seen that the object sought can be best attained by confining our attention as proposed.

CHAPTER II.

GENERAL REMARKS ON HARBORS, AND CERTAIN SPECIFIC WORKS.

It should be a matter of pride to all Americans, that immediately after the close of the great civil struggle, it was one of the first actions of Congress, to divert the tremendous energy then at work, from the paths of war to conquests of peace.

In every way possible, was enterprise invited to aid in the development of the country; by grants of land for the building of railroads; by the loan of the Nation's credit to the Pacific railroads; by grants of land for the establishment of agricultural colleges; and by the improvement of rivers and harbors.

As early as in June of 1866, bills were passed inaugurating work in continuation of plans begun before the war, and in beginning similar improvements at new points. It will be interesting to note where such work was to be done.

Upon the seaboard: Portland Harbor, Kennebec and Saco Rivers, Maine; Boston and Provincetown Harbors, Mass.; the Hudson River, and Delaware Breakwater; Susquehanna River; and Baltimore Harbor. Upon the great lakes: Michigan City Harbor, the only refuge at the lower end of Lake Michigan; Chicago, Milwaukee, Racine, Grand Haven, and seven smaller harbors on Lake Michigan, all engaged in lumber and grain trade; St. Mary's River and St. Clair Flats, of the connecting links; Saginaw River, the great lumber centre; Buffalo, Erie, Cleveland, Toledo, and seven other harbors on Lake Erie; Oswego, and three harbors on Lake Ontario; the great obstructions on the Mississippi River, Rock Island Rapids, and the Des Moines; the bar at the mouth of the Mississippi River; two general items for the great central rivers, the Mississippi, Missouri, Arkansas, and Ohio; and finally the Willamette River, of the Pacific Coast. For construction and surveys these bills aggregated $3,796,339.09.

In 1867 the bill was for $4,896,291.70, and began work on twenty-one new harbors, principally upon the Lakes, and eight rivers; and omitted fourteen items already provided for the preceding year.

Amidst the political excitements of 1868, and 1869, although detailed bills were not passed, yet in 1868 $1,500,000, and in 1869 $2,000,000 were granted for the preservation and improvement of rivers and harbors, to be allotted and expended under the direction of the Secretary of War, with certain general restrictions as to the selection of localities. An analysis of the allotments made under these acts, shows that the following works of national importance received the amounts set opposite:

Saint Clair Flats Ship Canal.,...................$228,560
Saint Mary's River.............................. 10,692
Rock Island Rapids, Mississippi River.......... 289,650
Des Moines Rapids, Mississippi River............. 478,200
Mouth of the Mississippi River............ 135,181
Falls of the Ohio, Louisville and Portland
    Canal 263,200

These six items cover 40 per cent. of whole, or $ 1,405,483

Besides these, the following important harbors were allotted:

Buffalo Harbor, N. Y............................. $ 89,100
New York Harbor (East River)................ 263,200
Boston 149,920
Baltimore 43,730
San Francisco....................................... 79,927

These five items, covering 18 per cent., were... $ 625,877

Thus, eleven items, out of fifty one to which sums were allotted in carrying out these acts, and these, evidently of national importance, received nearly 58 per cent. of the whole; while rivers and harbors of minor, but general importance, received amounts sufficient to do what was most necessary and desirable.

The river and harbor bill of July 11th, 1870, contained thirty-four items relating to harbors, and fifty-five to rivers, and amounted to $3,891,000.00; and it may be noted that ten of the eleven items last specified received forty-three per cent. of the total of the bill.

At this time a well devised system of operation was in progress, under plans understood by Congress, and in furtherance of projects selected by that body. The harbors upon which work was progressing were mostly upon the Great Lakes. The magnitude of the commerce connected with these harbors is shown in a report of a board of engineers, made in December 1873. In this, statistics are given of the number of vessels passing Detroit, and of all navigation interests throughout the country. From this we find that during the eight months season of navigation, 27,000 vessels passed Detroit. During the months of July and October 1872 the total entrances and clearances at New York were 2,333 with a tonnage of 2,126,000; while during the same months 9,220 vessels passed the St. Clair Flats, with a tonnage of 3,200,000.

During July 1872, the total tonnage of the United States, foreign and coastwise, entered and cleared, was 9,177,000; and of this 5,122,000 tons were on the lakes. By including similar statistics for the entire year it is shown that upon the basis of tonnage of entrances and clearances, the commerce of the Great Lakes was about forty per cent. of the total foreign and coastwise trade of the United States at that time.

The wisdom of fostering and developing this commerce cannot then be questioned; let us glance at the engineering features of the system. In general, at the beginning of this work, harbors were formed, or extended, by building parallel piers of crib work, filled with stone, and resting upon secure foundations; these piers extending far enough to reach deep water, and thus to protect the entrances of natural or artificial harbors from being filled up by sand driven along the shore by the waves. By dredging out the channel to the harbor formed by a stream or a land locked bay the whole became safe at all times, and accessible except in the most severe storms, coming from the most dangerous quarter.

But navigation interests soon called for more than that, and harbors at the ends of the lakes could not be entered when the wind blew directly into them, and the lee coast found at the *cul de sac* thus formed, was most dangerous.

At Buffalo, on Lake Erie, and Michigan City on Lake Michigan, other construction was demanded.

In 1839 Captain Williams urged the building of a detached breakwater at Buffalo; in 1866 Captain Tardy repeated the recommendation, and Colonel Cram in 1866 and 1867.

In 1868 a board of engineers adopted the plan, and made a location and estimate for a breakwater to be built of cribs fifty feet long, by thirty-four feet base, and from twenty-nine to thirty-seven feet high, to be begun in water of about twenty-five feet depth, and resting in a dredged trench. Construction began in 1869 and in two years the breakwater being over 1,000 feet in length, began to be of service. The difficulties of construction and the magnitude of the work, and the exposed location, make this of more than common importance. On June 30th, 1884, 5,691 feet of this breakwater were done, or under contract.

At Michigan City, although a board of engineers had recommended a breakwater in 1857, it was not until a subsequent board in 1870 proposed to build an outer harbor with three breakwaters, that Congress adopted the project, and work began in 1872.

At Chicago the harbor facilities called for enlargement, and an outer harbor was proposed in 1880; and the plan was adopted by a board of engineers and the breakwater began in that year.

At Marquette, on Lake Superior, an open roadstead; exposed to northerly and easterly storms; a crib breakwater was necessary from the first, and was begun in 1867.

At Cleveland, a board of engineers recommended, in 1875, a closed breakwater with an area of two hundred acres, for an outer harbor, to be formed by pile and crib piers extending to twenty-seven feet depth of water, with crib work parallel to the shore for the outer breakwater. The prolongation of an existing pier to within 300' of the lake front crib work, would complete the harbor. Work authorized by Congress in 1876, and construction at once begun. At the close of the fiscal year 1884 the shore arm was 3,130 feet long, and the lake arm 4,030 feet, with a spur one hundred feet long on the north side near the end of the lake arm. The outer breakwater is placed in water having an original depth of about five fathoms.

At Oswego, on Lake Ontario, a board of engineers recommended, in 1870, that a new outer harbor be built, as existing harbor works were insufficient. Work began the following year, and was finished in 1883 to the following extent: The west breakwater is a shore arm about 900 feet long, running northerly, and the lake arm about 4,900 feet long, running from the outer extremity of the shore arm about north 60° east, terminated at its eastern end by a wing running shoreward about two hundred and fifty feet.

This work is peculiarly subject to storms, and it is noted that waves eighteen and one-half feet above the normal surface of the lake have been measured. Nothing but the most substantial structure could stand under such storms.

The necessity for an artificial harbor of refuge between the head of the St. Clair River and Saginaw Bay, was recognized as early as 1870, there being no natural harbor in that distance. After many projects and examinations, a board of engineers finally selected Sand Beach as a proper locality, in 1873, and adopted a plan and submitted estimates. Work was begun in 1874, in accordance with these plans, and by June 30th, 1875, 3,070 linear feet of breakwater were in place, to the immediate advantage of commerce. On June 30th, 1884, the total length of breakwater in place was 8,130 linear feet, partly unfinished. Within the harbor area large dredging has been necessary to complete the plan. During that fiscal year 1,182 vessels took refuge in the harbor.

It is not within the province of this essay to point out all the works executed by the United States, through officers of the Engineer Corps, upon the lakes, so that only allusion can be made to such works as Duluth, piers and dredging; Eagle Harbor, rock excavation; Saginaw River, and Toledo Harbor, extensive dredging; and many minor works; but it is intended to call attention, here, to the fact that no candid and just opinion can be formed of the scope of river and harbor improvement bills, and especially of the connection of officers of the Engineer Corps therewith, without a critical examination of the records and results upon the Great Lakes.

At the entrance of the St. Clair River into Lake St. Clair, a narrow and tortuous channel, without sufficient depth, had greatly hampered navigation. This, known as the St. Clair Flats, was improved in 1855-8, and in 1866, by dredging, under three appropriations amounting to $145,000. In 1866 Colonel Cram proposed to construct a straight canal with dikes, built of piles and sheet piles, and filled with the dredged material. Between these dikes, 8,200 feet in length, was to be a dredged channel three hundred feet wide and thirteen feet deep. The plan was approved, and carried out by contract begun in 1867 and finished in 1870. By thus giving a straight channel, with sufficient depth, this canal gave great relief to the entire commerce passing between Lakes Huron and Erie. The total cost was somewhat less than $400,000. In 1872, to meet the increasing draft of vessels, it was proposed by Major Poe to deepen the canal by dredging an interior channel two hundred feet wide and sixteen feet deep. This was done, by contract in 1873 and 1874, at an additional expense of about $100,000, which included necessary repairs. The care and maintenance of the canal involves a certain small annual sum. In construction the semi-fluid character of the sand dredged made extra care and expense necessary.

I propose to class together, for a brief general review, the great obstructions to navigation found at the Falls of the Ohio; at the Muscle Shoals of the Tennessee; at the Sault Ste Marie; at the Des Moines Rapids, and the Rock Island Rapids, of the Mississippi. At all but the last mentioned, ship canals have been required; and at the Sault Ste Marie no navigation was possible without a canal. At the Falls of the Ohio, as early as in 1793, a canal was discussed. In 1825 Kentucky granted a charter to a stock company to construct a canal around the Falls within the State; and by acts of Congress in 1826, and 1829, the United States bought shares in the company to the amount of $233,500. The canal was first opened for business in December 1830. Within eleven years tolls had been collected greater than the original cost of the work. By a series of enactments the State endeavored to apply the tolls to the purchase of the shares of private parties, and then turn the whole over to the United States, that it might be run in the interests of commerce, without unnecessary expense. By 1855 this was done, and the general wish was, that the United States would assume the canal and enlarge it to accommodate increased demands. The original company was, however, obliged to begin the enlargement itself, and until forced to suspend the work in 1866 through lack of means, continued their efforts to meet the urgent demands of commerce. The United States first took an active interest in the matter in 1867, by ordering a survey as to needed works. Major Weitzel, as a result of the survey, proposed to continue the work of enlargement then in progress under the board of directors, and construction was begun in 1868 under an allotment from the bill of that year. In 1870 an adequate sum was granted, and by February 26th, 1872, the new locks were first opened to commerce; and in November 1873, the work of enlargement was virtually completed. It was not until June 10th, 1874, however, that the United States took control of the canal by virtue of the requisite legislation. During this time the State of Kentucky and the board of directors of the canal did all in their power to facilitate the transfer.

The locks are combined, two in number, each three hundred and seventy-two feet between mitre posts, with an available length of three hundred and thirty-five feet; they are eighty feet wide, and have lifts of twelve and fourteen feet. The total cost of this work to the Government, from 1826 up to 1882, deducting tolls received, was a little over $2,600,000. The canal was made free from tolls July 1st, 1880. When the stage of water on the Ohio is high, vessels pass up and down the Falls without using the canal. Statistics of its commerce will be found later on in connection with the Ohio.

[1]The State of Alabama began, in 1831, a canal designed to overcome the Muscle Shoals of the Tennessee River. Congress aided the project by a land grant of 400,000 acres. The canal was not finished. The United States undertook its reconstruction and enlargement in 1875. This canal is not yet complete. A description will be found later on.

The impassable falls, known as the Sault Ste Marie, in the Saint Mary's River, barred all navigation between Lakes Superior and Huron. As commerce increased the obstruction became more of a burden, until in 1852 Congress gave a land grant of 750,000 acres to aid in the building of a canal. In 1853 the State of Michigan accepted the offer, and through a private company built the canal, which was opened to commerce on June 15th, 1855. This was about the first ship canal in the United States. The locks and gates were the largest made in the country up to that time, and the depth of water greater than given to any prior work. The locks were two, and combined, each three hundred and fifty by seventy feet, with a lift of nine feet each; the depth on the mitre sills being twelve feet. By the year 1870 these dimensions no longer sufficed for the rapidly growing commerce, and the United States assumed the work of enlargement. A new lock was built parallel to the old ones. This lock is 515 feet long and eighty feet wide, interior dimensions, with a

---

1. 1868, p. 776.

lift of eighteen feet, and has a depth of sixteen feet on the mitre sills. The width at the gates is sixty feet. The old canal was deepened, and widened so as to include both the old and new locks. The guard gates at the head were removed and new ones placed 700 feet above the old location, and at a lower level. Dredging in the river below the canal was necessary also. Major Poe began, and Major Weitzel finished the work of enlargement. In eleven different appropriations $2,405,000 were granted for this purpose, and the new lock was opened to navigation September 1st, 1881. The canal was transferred to the United States, by the State of Michigan, by its act of March 3d, 1881. The officer in charge at the time of completion, Major Weitzel, called attention to the great growth of its commerce, and of additional facilities needed to accommodate it. The tonnage passing over that portion of the river was reported thus:

From 1867 to 1871 inclusive................................2,957,173
"   1872 to 1876   " 5,991,247
"   1877 to 1881   " 8,521,332

AFTER THE OPENING OF THE NEW CANAL

	1881.	1882.	1883.	1884.
Vessels, number..................	3,304	4,676	4,163	4,768
Lockages, number...............		2,618	2,330	2,569
Tonnage, registered.............	1,802,571	2,379,216	2,130,181	2,333,257
Tonnage, freight.................	1,258,468	1,878,154	1,874,404	2,540,799
Coal, tons........	195,448	385,204	464,320	764,915
Copper, tons....................	21,189	30,546	25,937	33,536
Iron Ore, tons..................	653,518	910,964	804,766	933,107
Pig and other Iron, tons......	53,935	104,664	78,909	93,103
Grain, bushels..................	3,901,332	3,699,268	5,068,417	7,490,938
Flour, barrels..................	525,060	536,637	427,866	891,291

The increase in tonnage during the year 1884, (season of navigation), is equal to the entire commerce through the canal for the first five years after it was opened in 1855. The great increase in coal going up, and grain and flour coming down, is noticeable, and the lumber interests are but beginning the growth which in time will be very great. In 1884 122,000,000 feet, board measure, passed the canal. Should the rate of increase be maintained the canal will be blockaded within four years. Special reports, urging additional enlargement, have been made. As commercial investments the United States never made better than this, and the Louisville and Portland Canal; and as an engineering example, nothing in the world can be quoted superior to the Sault Ste Marie lock.

At the Des Moines Rapids, of the Mississippi River, a first appropriation of $100,000 in 1852, and a second in 1856 of $200,000, gave no practical relief. In 1866 the sum of $200,000 was again given, and after surveys and various projects a board of engineers in 1866 selected the plan presented by Lieut.-Col. J. H. Wilson, for a canal on the west shore, from Keokuk to Nashville, seven and six-tenth miles, to overcome the lower fall. This canal is two hundred and fifty feet wide with a navigable depth of five feet, and has two lifts and one guard lock. The locks are three hundred and fifty by eighty feet, interior dimensions, with lifts of eight, and ten and one-third feet; and the guard lock at the head of the canal is arranged to act as a lift lock if necessary. After an interval of four miles, the upper chain is reached, and through this a channel is excavated in the solid rock two hundred feet wide; and of the same, five feet depth. Through the eleven and one-half miles of the rapids, the total fall is twenty-two feet, nearly. To prosecute to advantage, from the beginning, large sums were needed, as construction could be carried on simultaneously along the whole line, and delays from existing traffic could not arise. When begun, no work in charge of officers of the Engineer Corps was of equal magnitude, and although begun later, the two canals of which enlargement was made, were completed years before the Des Moines Canal. It was repeatedly urged that adequate sums be given, but it was deemed best by Congress to limit the amounts available at any one time, and delays were necessary, and expensive. Thus the canal, on which the first contract was let in 1868, was not opened for business until August 22d, 1877; and as yet it is not complete in accordance with the plans.

At the time of opening, the amount appropriated was $4,281,000, and $100,000 was estimated as still needed. Although like the Falls of the Ohio, the Des Moines Rapids are passable at suitable stages of water, they are not navigable at low stages, and these occur at the busy season; thus causing greater relative loss to navigation than on the Ohio, where the winter season, with plenty of water and freedom from ice, gives advantages not possessed by the upper Mississippi.

Since the date of opening, the canal has been operated under the charge of the Engineer Department. The statistics of its commerce will be considered in connection with the upper Mississippi. Colonel Macomb succeeded Lieutenant-Colonel Wilson in charge, and Captain Stickney was in local charge from 1872 to 1878, when he relieved Colonel Macomb.

The Rock Island Rapids extend from the foot of the island of the same name, a distance of fourteen and one-half miles up the river. Throughout this distance the bed of the river is of limestone, and although the total fall is only twenty-one and one-half feet, or a little over eighteen inches to the mile, yet beds, or reefs of rock, occurring at seven different localities, make serious obstructions, dangerous themselves and having a much greater fall in short distances. These chains, as they are termed, projected from each shore, and nearly meeting or overlapping, made narrow, tortuous channels, with swift current, or completely block the river, as at Moline chain, where before improvement only thirty inches of water was found in the channel. Sycamore chain was the worst place on the rapids, owing to the sinuous channel and great velocity. Between these chains, pools, with good, navigable channels, of sufficient depths, existed, and eleven miles of good navigation, to three of bad, made up the rapids. During the season of navigation, no less than one-third of the time it was either impossible or dangerous to pass these rapids, and as early as 1852 it was thought to begin work upon them.

An appropriation of $100,000, in 1866, began the work in earnest. It was proposed by Lieut.-Col. J. H. Wilson, in 1866, to cut through the limestone, a navigable channel, two hundred feet wide, and four feet deep; and this project was approved by a board of engineers. Work began in 1867 under contract, and was thereafter carried on partly by contract, and partly by hired labor under the United States.

In some places coffer dams were used to advantage, and in others the rock was cut by heavy chisels operated by machinery from boats. The rock thus cut in position was removed as was most convenient. Drills were used from platforms, supported by temporary tripods resting on the bed; and the rock blasted under water. By the end of 1875 the work was completed, in general, through the principal chains. The amounts excavated, increased from 5,508 cubic yards in 1867, removed from Duck Creek chain, to 23,788 cubic yards in 1869, from the same, Moline, Sycamore, and Campbell's chains. The total excavated by the end of 1875 was 84,358 cubic yards. The amount appropriated up to that time was about $1,090,000. Subsequently, scattered patches of rock were removed, the channel rectified in places, buoys were set and maintained, and loose rock, snags, and boulders taken from the channel when brought into it by floods or otherwise. This work cost $77,000 during the past ten years. Colonel Macomb relieved Lieutenant-Colonel Wilson, and remained in charge of the work until 1877.

The railroad bridge crossing the river just below this rapids was decided, in 1859, to be an obstruction to navigation, and subsequently legislation led to its rebuilding, partly at the expense of the United States. The piers and abutments were begun, under contract, in 1868, subsequently completed by hired labor under the Engineer Department; and the superstructure was built by contract. The whole was completed in 1873. Although under the control of the Engineer Department during construction it does not come within the intention of this work to more than allude to it.

From a general review of the history of the great works just outlined, we find that the genius and instincts of our people are such, that it is hardly practicable to rely upon private, or state action, in the matter of great obstacles to navigation. While offering, at times, privileges to private companies, it seems natural to look to the general government for final action and control, and to invite an assumption of the burden, and jurisdiction by the United States, in behalf of the general interests of the various States. No question or doubt has ever arisen as to the expediency of undertaking the improvements indicated. That the work has not been done as promptly as might be desirable, in some cases, but shows the caution of Congress — not its wastefulness. If it were the intention simply to show herein the skill and engineering talent exhibited during the past eighteen years, it would be best to rest with full descriptions of these works, but it can only be said, here, that in planning future, or continuing present systems, a just idea cannot be obtained without a due study of what has already been done.

Of the harbors upon the Atlantic, upon which work was resumed, or begun, in 1866 and 1867, the one in which the results are most distinctly successful, is Baltimore.

In 1852, under an act of the State of Maryland, the citizens of Baltimore began to deepen the approach to the city. The channel at that time had a minimum depth of sixteen feet. The United States made an appropriation of $20,000 in 1852, and one of $100,000 in 1856. Under an agreement between Captain Brewerton, of the Engineers, and the board of commissioners for the harbor, a channel was selected which extended in a direct line from Fort McHenry for six miles, then, with a change of direction, for nine miles further in a direct line, to deep water. It was proposed to secure, by dredging, a channel one hundred and fifty feet wide, and twenty-two feet deep at mean low water. When available funds were exhausted, this work was not complete. In 1866 it was resumed, with slight change of plan, and a proposed increase in width to two hundred feet, under charge of Major Craighill, of the Engineers, who carried the project to completion, and remains in charge. In view of the rapid increase in commerce, the project was revised in 1872, with the intention of increasing the width to two hundred and fifty feet, and the depth to twenty-four feet. This channel was secured in 1874, and has since been maintained. About this time a wonderful growth in the foreign commerce of Baltimore set in, which in eight years doubled the number of arrivals of vessels from foreign ports, and quadrupled the value of exports from the port. In 1881, therefore, it was proposed to increase the depth throughout to twenty-seven feet, and on a part of the route cut a new channel which would be a cut-off, and probably somewhat more stable. In 1884 Lieutenant-Colonel Craighill reported that a narrow channel, twenty-seven feet deep, existed throughout, and is doubtless widened and completed by this time; probably furnishing the harbor all the facilities needed for the present, if maintained through repairs, which in such cases are always necessary. The entire dredged channel, making but three changes in direction throughout its length of more than fifteen miles, is easily navigable. The influence of the improvement upon the commerce of Baltimore is marked. In 1866 the import and export returns for the year are reported at $15,780,790 in value; and in 1869 the arrivals from foreign ports 650; value of imports $21,017,313, and of exports $12,460,687. These figures steadily increase to 1876; when by rapid increase a maximum was attained in 1880. During this fiscal year the arrivals from foreign ports numbered 1,794, and the value of exports amounted to $77,125,022. The total tonnage of all entrances and clearances, foreign and coastwise, was 5,518,453. Since 1877 Baltimore has been the second city on the Atlantic in receipts and exports of wheat, and between 1873 and 1877 passed Philadelphia and Montreal in that respect. Since 1881, from causes which seem to be shared by other Atlantic seaports, the exports have fallen off, as well as the arrivals from foreign ports. During the interval from 1869 to 1884 the value of imports have varied from a maximum of $28,836,305 in 1872, to a minimum of $11,423,665 in 1884. The total amount appropriated for the harbor of Baltimore, and allotted it since 1866, has been $2,930,830. Since improvements began the city and State have expended about $484,000 on the harbor.

The completion of the breakwaters at Portland Harbor, Maine, and in Delaware Bay; the preservation of Boston Harbor, and the removal of Tower and Corwin Rock, Kelley's and other ledges in that harbor; and the continuation of the improvement of the Hudson River, upon which corporate, State and United States work had been done for many years, were among the first objects of the bills of 1866 and 1867. Beginning with 1868, the great work of the removal of the submerged rocks and reefs of East River, New York Harbor, was inaugurated, upon which, alone, a volume could be written.

Ledges of rock and other rock shoals, or reefs, at varying depths, were subsequently removed from the Kennebec River, Machias River, Narragaugus River, Penobscot, and Sullivan's River, Maine; in Portsmouth Harbor, N. H.; Passaic River, N. Y.; the James River, Va.; and notably Blossom Rock and Noon Day Rock, in San Francisco Harbor. Breakwaters and tidal jetties have been built at many points, at different harbors on the Atlantic, and notably at Charleston, S. C., and at Wilmington, California.

The peculiar and excessively difficult work at the mouth of the Cape Fear River, N. C.; and the tedious and extensive dredging operations upon the Delaware River, and in Philadelphia Harbor; and in and approaching Mobile Harbor, Ala., are of great magnitude; and those at Oakland Harbor, California, are phenomenal.

The extensive, almost universal use of dredging at localities too numerous to mention, both by contract and with machines which for special cases have been devised or adapted by Engineer officers, needs no special description or elaboration.

Official reports are full of the literature applicable to harbor improvement. The principles which govern the treatment of tidal harbors, are fully explained and discussed in the reports on Boston, Charleston, S. C., the mouth of the St. John's River, Fla., San Francisco and Wilmington, Cal., and the much vexed question of Galveston Harbor.

If it be seriously entertained, as it has been lightly charged by some in important positions, that officers of the Corps of Engineers have not shown themselves able to cope with engineering problems coming before them, it would be well to examine the history of what has been done, before committing one's self to such an opinion.

2

The records will show, as the brief outline just given does, that the work done or required at the various harbors of the United States, has been varied in the extreme; and that engineering principles and devices have been applied in great, and in minor instances, with general success. Unless the history of the past eighteen years be entirely ignored, careful examination, and fair judgment, will decide that little reason exists ·to doubt the wisdom of Congress in these matters, or the ability of the Corps of Engineers in their execution.

CHAPTER III.

THE OHIO RIVER, IN ITS UPPER PORTION, AND SOME OF ITS TRIBUTARIES.

It will be best to begin our investigation with the Ohio River, for many reasons. These are, its vast commerce, which, beginning with the very occupation of the country, and growing with steam in its development for navigation purposes, has been a great factor in the settlement of the Central States; and again, the character of the river and its tributaries, rising in mountains rich in timber and mineral resources, flowing down steep slopes, and with stable beds and banks, subject to great and sudden floods; and yet, with all this, capable of definite improvement, and permanent artificial channels. A region of large area, and with a generous rainfall, yet at low water season with but the smallest fraction of a discharge in the water courses, compared with flood outflows. With all this, a vastly diversified condition of agriculture, mines, timber and manufacturing interests, throughout the region and necessarily to be considered in each case; thus making a commercial kaleidoscope, not only of the whole, but even of different parts of the same tributary. At the risk of offering too much detail I shall treat this portion of the navigable interior system quite fully.

From the junction of the Alleghany and Monongahela Rivers, which form the Ohio, to the junction with the Mississippi at Cairo, is a distance of 967 miles. From Pittsburgh to Beaver, a distance of twenty-six miles, the average fall is about fifteen inches to the mile; and from Beaver to Wheeling, sixty-four miles, the average is about nine inches per mile.[1] In this stretch the steepest natural low water slopes in the river are found to be at:[2]

Horsetail,	5	miles below Pittsburgh, one foot fall in							461
Deadman's Island	14	"	"	"	"	"	"	"	513
Twin Islands,	85	"	"	"	"	"	"	"	781
The Trap,	11	"	"	"	"	"	"	"	800

1. 1881, III, p. 1929.　　2. 1875, II, p. 609.

At the lowest stage of the river the channel is confined, at the last mentioned place, to a width of but two hundred and twenty-five feet; and there is often but eighteen inches, and at times but twelve inches for navigation.[3] The quantity of water during low water is small, and as gauged at Pittsburgh in 1867-8 and 9, is reported as follows:

At zero of the gauge, 1.666 cubic feet per second.
" 3" " " " 3.000 " " " "
" 6" " " " 4.387 " " " "
" 9" " " " 5.810 " " " "
" 1' " " " 7.274 " " " "
" 2' " " " 13.554 " " " "

The low water discharge was calculated in 1838 to be 1661 cubic feet per second. No tributary of importance joins the Ohio until the Muskingum, at one hundred and seventy-one miles below Pittsburgh, comes in. The Little Kanawha River, at one hundred and eighty-three, and the Hocking, at one hundred and ninety-seven and one-half, are the only remaining ones of note until the Kanawha River, at two hundred and sixty-three, joins. From Wheeling to Letarts Falls the slope is 7.42 inches per mile, and the distance one hundred and forty-one miles. Between Pittsburgh and Wheeling the bed is of coarse gravel and boulders, and sometimes rock. Below Wheeling the gravel becomes finer, the boulders fewer, and bars of river sand appear. The width between banks does not increase much from Pittsburgh to Cincinnati. Many large islands are found in the upper part of the river, and forty-five, varying in size from a few acres to 1,900 in extent, lie between Pittsburgh and the mouth of the Kanawha; many are cultivated, and all are permanent in position. At every island the navigation is more or less injuriously affected, and at the mouth of every tributary lies a bar of greater or less magnitude.[4] The majority of these do not affect the navigation materially; others cause considerable trouble in certain stages of the river; some remain the same, or very nearly the same, from year to year; some change more or less with every flood; some are wearing away, while others are increasing. Some of the bars are of gravel, some of sand, light or heavy, while some of the shoals, as at Beaver and Letart's Falls, have solid rock bottoms. In some places rocky shoals project from the shore. Nearly all the ripples are caused by natural gravel dams, in some cases underlaid with rock. The whole river, is, in fact, a succession of natural dams and pools. In the upper part of the river the bed is unyielding, and the slopes are fixed. During the earlier stages of improvement of the river, it was the practice to preserve these pools and concentrate the flow of water by closing many island chutes, and running out in other places dikes, to force the water into the channel. Operations began in this way in 1837, and

---

3. 1873, p. 499.　　4. W. M. Roberts' reports.

were carried on until 1844 in many places. After that time nothing was done until 1866, when, upon resuming operations, repairs were made upon the old work, and some of the unfinished plans were carried out. A survey of the river was begun in 1838 and carried down stream. The following table embodies much information thereby collected, and indicates points where contraction works were begun, and carried on during both intervals. Some other data have been added, and the whole is intended to cover the principal characteristics of the river as far as the mouth of the Kanawha River.

	Miles from Pittsburgh	Depth of Channel at low wat'r	Low water discharge, cu. ft. per. sec. 1888.	Low water velocity, miles per hour.	Length of Channel.	Fall during same.	Low water width of Channel.	Width between banks.	Flood height.	Areas of cross section for discharge.
		In.			feet.	feet.	feet.	feet.	feet.	
Alleghany City...........	2							1,100	35.6	
Glass House Ripple *†....	2	20	1,661	1.11	550	2.25	137			
Horse Tail Ripple* or Davis Island................	5½	18		2.11	1,280	2.43				
Merriman's Ripple.......	10¼	20		2.00	2,900	1.86	250			
White's Ripple *†	11	26		3.00	1,738	2.64				
Wollery's Trap*†	11¾	16		2.51	2,300	2.41				
Deadman's Ripple*.......	14¾	18		3.21	3,250	4.41				
Logtown Bar *†..........	18¾									
Beaver Shoals*†..........	26¾	24								
Line Island.............	41	24		4.21	400	1.20	93			
Brown's Island *†........	60¾									
Wells' Bar.............	68	24		1.81	929	1.15			45.0	
Twin Island *†..........	84¼	31			3,250	.48				
" " Lower chute.		18		2.51	2,400	1.21				
Wheeling Island, head †. {	88½	22	[1868] 3,000 at 14" stage	2.21	7,725	2.66		900		
" " 'at foot...		18		1.91	5,700	1.79			52.08	
Big Grave Creek Bar......	102	30		3.81	1,587	1.18	420			
Captina Island *†........	106¾	18			7,200	3.65	270			
Fish Creek *†............	112	18		2.11	3,150	2.12	203			
Grand View Shoals †......	143	18			7,325	.43				
Sheet's Ripple †..........	146	18			7,280	.43				
Petticoat Bar and Ripple*†	147	15			4,300	1.40				
Middle Island............	152	21	2,487		2,700	1.80				
Carpenter's Bar..........	166	18			3,530	1.52				
Marietta Island, head †.. {	168	36	gauge [12'] [1871] 35,503		4,650	.66			52.09	
Mouth Muskingum River..	171									
Muskingum Island, head †.	174	14			3,020	1.38				
" " foot...	176	20			4,750	1.41				
Cole's Island *...........	180	21	2,558	2.81	3,000	2.47	135			
Parkersburg........... } Mouth Little Kanawha R. }	183							1,200	43.5	
Blennerhassett's Island *†.	184½	30	2,224	2.00	2,150	1.19	378			869.4
" " chute		12	313	1.81	1,430	.61	77			100.1
Mouth Hocking River.....	197½									
Belleville Bar...........	200	18		2.51	7,500	2.22	188			362.1
Belleville Island †........	202	18	2,700	2.91	5,000	4.60	298			685.4
Buffington Island, left *†..	214	18	2,100	1.61	4,080	1.14	741			750.2
" " right...		22	588	1.61	2,250	1.11	167			250.5
Sand Creek Bar..........	219½	20		3.00	2,300	3.65				
Letart's Island, head *.....	232	17		2.21	712	.47				
Letart's Falls.............	234	24		3.81	2,800	3.15				
Mouth of Kanawha River.	263									

* Dams or dikes built or begun during the period 1837-1844.

† Repairs made to dams or new ones built 1867-1884.

Gaugings of the upper river are not reported with the fullness that the published surveys of some other rivers exhibit. The Board of Engineers upon the Ohio River Bridges report (1871 p. 397) on April 19, 1871, as the result of their surveys, the following, on this subject:

	Miles below Pittsburgh.	Gauge Reading.	Area Cross Sect. sq. ft.	Discharge cu. ft. per s.	Vel. Miles per hour.	AREAS. High Water sq. ft.	Low Water sq. ft.
Steubenville	67.	9.′	....	37.858	2.65	77.189	3.179
Bellaire......	94.	8.5	11.105	36.261	2.226	74.787	3.425
Parkersburg	183.	8.′	15.676	41.058	1.75	114.854	7.271
Cincinnati ...	467.	6.′	10.875	30.240	1.76	80.475	6.526

During the interval from 1844 to 1866 the river was left to take care of itself, and it was remarked that the charts of the first survey showed that the stream itself had not essentially changed during this interval. From the continuous record of the gauge kept at Pittsburgh since 1855, it is reported that between 1855 and 1878, the gauge stood at subjoined readings for an average, as follows: [5]

Gauge read above 2′ and below 3′,   37. days.
      "   "   3′   "   4′,   46.8   "
      "   "   4′   "   5′,   43.5   "
      "   "   5′   "   6′,   40.4   "
      "   "   6′   "   7′,   36.8   "
      "   "   7 feet,   118.4   "

From this record of 24 years the average duration in days of a stage allowing a 3-foot navigation was 285, average duration in days of a stage allowing a 6-foot navigation was 155. In no month of the year can navigation be called absolutely continuous; it is more or less intermittent in every month. The best boating season in the vicinity of Pittsburgh, (when not interrupted by ice), and throughout the whole Ohio, is from the beginning of December to the end of May, but the duration of navigable stages increases as we descend the river. A study of the gauge record indicates it possible that the periods of navigation are becoming shorter, but it would be well to investigate the question further, and from similar records kept at other points.

When the work of improvement was resumed in 1866, it was conducted by Mr. W. Milnor Roberts, civil engineer, in continuation of former plans, and beginning at Pittsburgh, works of contraction were carried on at the worst places, with as much speed as the funds allowed; repairing the old works and inaugurating new. In addition to this, dredging was resorted to, and the removal of rocks, snags, and wrecks. While carrying out this work Mr. Roberts studied the requirements of the river, and reported, in 1870, that slack water nav-

---

5.   1879, II. p. 1310.

igation only, would give the depth of water required. He proposed using locks and dams with chutes for high river stage exceeding six feet. His estimate to secure that depth from Pittsburgh to Louisville was $17,052,207. Maj. W. E. Merrill, Corps of Engineers, was placed in charge, in June, 1870, and has remained in charge since then. Operations were continued as before, for some years, one important change being, the building and operation, by the United States, of a dredge, the "Ohio," and necessary dump boats, in 1871 ; and a second, the "Oswego," in 1873. Dikes were built at Medoc Bar, 485, Rising Sun, 502, Warsaw, 524, French Island, 760, and Evansville 783 miles below Pittsburgh; all below Cincinnati, and all works of considerable magnitude; during the years 1871 to 1874. In 1872, Colonel Merrill stated[6] that after long trial of other methods the commercial interests of the upper end of the river had about concluded that in that part of the river the only radical method of improvement was the construction of locks and dams, the usefulness of these depending, however, upon the practicability of making chutes two hundred to three hundred feet wide, with gates and inclines so constructed that coal flats could be safely passed from one pool to another without being divided. On April 16th, 1872, a board of engineers [7]consisting of Maj. G. Weitzel, (then in charge of the improvement of the Louisville and Portland Canal), and Major Merrill, was appointed to examine and report upon the plan of Mr. F. R. Brunot for movable hydraulic gates for chutes and locks, its applicability to the improvement of the Ohio and other rivers, and an estimate of the cost of construction; and to this duty was added, May 1st, 1872, the general consideration of the whole subject of movable hydraulic gates, and of all other proposed plans for this object. This board reported January 31st, 1874. After giving a full description of movable dams found in France, and sluices in India, twelve methods in all, the board examined certain other devices brought before it in the shape of models and plans, and in a final recommendation proposed to experiment with a navigable chute, to be opened and closed by a hydraulic gate, in one of the dams of the Monongahela Navigation Company, with a view to the adoption of the system on the upper Ohio, if it should be found practicable. This proposition, however, was never carried out.

The United States dredges have done much, and varied work at all points along the Ohio, removing parts, or all of bad bars, rocky points, obstructions, and wrecks, besides assisting in the works of construction at times. The Ohio, beginning work in 1873, was given an iron hull in 1880; the Oswego, built in 1874, was given an iron hull in 1883. The record of the cubic yards of material, gravel and rock, removed by both, in successive years, is here given:

---

6. 1872, p. 399.   7. 1874, I. p. 415.

1873, 29,380 cu. yds.	1880, 81,180 cu. yds. and 1 wreck
1874, 78,634 " "	1881, 70,695 " " " 3 "
1875, 85,769 " " and 3 wre'ks	1882, 47,671 " " " 2 "
1876, 142,092 " " " 9 "	1883, 110,771 " " " 4 "
1877, 114,270 " "	1884, laid up for lack of funds.
1878, 81,879 " "	

Snags were removed by contract, by government snag boats, and finally, in 1876, an iron hull, completely equipped, snag boat, the E. A. Woodruff, was built, and has since been operated continuously in the removal of all such obstructions. This boat is capable of handling any possible obstruction. In 1883, a sycamore tree measuring 7' 8" diameter at the butt, and weighing 364 tons, was removed; and an average weight of eighty of the largest snags removed that year was 92½ tons.

Upon September 1st, 1874, Colonel Merrill modified his views on the radical improvement of the river, and stated,[8] that after long study he had come to the conclusion that the best method of improving the Ohio, at least in the upper part of its course, was to follow the plans that had been successful on the Seine, Yonne, Marne, Meuse, and other French rivers. This method of improvement by movable dams on the Chanoine plan, would require, 1st, a movable dam of Chanoine wickets divided into navigable pass and weir; 2d, a lock large enough to pass an ordinary coal-fleet through in one lockage. Approximating the positions of such dams to secure a six foot navigation from Pittsburgh to Wheeling, thirteen were located, each with a six foot lift. The average length of the dam was given as 1,473 feet,

At an estimated cost of... $ 374,971
Estimating the cost of the lock with the dam.............. 574,971
The whole cost of the improvement from Pittsburgh to
    Wheeling will be................. 7,474,623

Recommending work at all points simultaneously, Colonel Merrill adds that there is no doubt, whatever, of the absolute necessity of using locks, in any rational plan, for improving the upper Ohio so as to secure a six foot navigation.

Meanwhile, the question of the Ohio below Louisville, was practically tested by long and substantial dikes built at French Island, 760 miles below Pittsburgh, to a length of 3,498 feet, at Evansville, 783 miles, and at Henderson, 797 miles below Pittsburgh. The first effect of the French Island dike was excellent and rivermen were delighted with it, thinking a radical improvement had been made.[8] Unfortunately the sands had reformed the bar just below the end of the dike, so that very little practical advantage had resulted from the work. An extension of 1,000 feet was added to the first 2,000, with the hope that the sand would be driven into deeper water and be so

---

8. l. c. p. 406.   8. 1875, I. p. 679.

distributed along the sides of the channel as to leave good navigable water in the middle. Henderson dike,[9] 1,900 feet long, and six feet above low water, seems to answer its purpose well, though a lack of water on the bar at the foot of the island, a bar not existing when the dike was built, may yet require a new dike to cause its removal. The greatest obstacle to the successful improvement of the lower Ohio, comes from the enormous masses of sand and gravel that travel down stream with every rise.[10] Behind the dikes at French and Henderson Islands, and at Evansville, sand bars formed as high as the dikes, and in places where before, or during their construction, there was deep water. At the Cumberland Island end of the Cumberland dam, where in 1873 there was about ten feet of water, in 1874 was ten feet of sand, showing a deposit of twenty feet in one season. This sand is mixed with gravel, which was carried over the dam. Evansville dike was lengthened in 1879, and all three strengthened and repaired in 1879 and 1880. Very substantial dikes built of crib work and filled with stone were subsequently built in 1882–4 at the following points, and consumed material as given—

	Miles below Pittsburgh	Length of dikes feet.	Timber, feet B M.	Stone, cubic yards.	Iron bolts and spikes. lbs.	Brush, cords.
Four Mile Bar..............	457	2,135	....	....	....	....
" " "	.....	2,477	....	....	....	....
Portland Bar...............	603	2,581	489,631	17,741	35,810	1,661
Puppy Creek Bar..........	743	2,935	365,887	32,583	31,274	13,663
Grand Chain...............	943	2,700 ⎱	....	....	....	....
" "	.....	3,232 ⎰	422,463	36,667	84,037	1,292

[11]Under date of February 25th, 1875, Colonel Merrill expressed himself in favor of extending the movable dam system throughout the entire river, qualifying his statement as to its applicability below the Falls of the Ohio, by saying that although not assured of its serviceability there, it was a better system than one of permanent dams, and the only other system promising six or seven feet of navigation at low water, the system of dikes not being likely to afford more than four feet at dead low water, and then only after an immense development of such works. He then gave an approximate location to sixty-eight dams which would be required to give six feet of navigation between Pittsburgh and Cairo. The total length of these dams would be 118,885 feet, including the locks, and 10,840 linear feet of dam across island chutes. An estimate based upon these figures gave for the total cost of such improvement $38,696,671. Considering additional difficulties likely to occur in the lower river

it was deemed best to give the total estimate as $40,000,000. The river and harbor bill of March 3d, 1875, appropriated $100,000, "to "be used for and applied towards the construction of a movable dam, "or a dam with adjustable gates, for the purpose of testing substan- "tially the best method of improving, permanently, the navigation of "the Ohio River and its tributaries."

A board of Engineer Officers,[12] consisting of Lieut.-Col. H. G. Wright, Majors G. Weitzel, O. M. Poe and W. E. Merrill, and 1st Lieut. F. A. Mahan, as recorder, reported under date of April 19th, 1875, that "with regard to the plan of this proposed dam, as sub- mitted by Major Merrill, the board are of the opinion that it is satisfactory. The dimensions and details for the lock have been care- fully matured, and are considered to be judicious. The site chosen for the dam is recommended to be adopted. The lock will be seventy-eight feet wide and have an available length of 630 feet." Considering the great width of the river to be closed the board thought Chanoine wickets should be used for the navigable pass. For the two wiers no system was recommended until the time of actual construction, when additional study and experience abroad might result in a better solution of this part of the problem. A careful estimate was made of this proposed movable dam and its lock at Davis Island, and given at $463,103. To compare this estimate with the cost of French locks and dams, the following table is given: [13]

---

12. 1876, II p. 25.   13. 1874, I. p. 447, 454.

RIVERS.	Number of Dams	Width of Lock feet.	Width of Pass. feet.	Length of Weir feet.	Average cost of Lock and Dam.	Total Cost
Yonne	7	34.44	115.29 to 98.24	96.60 to 81.80	$ 60,230	$421,610
"	8	34.44	115.95 to 115.29	257.15 to 165.48	87,780	702,240
"	7	34.44	115.29	165.48 to 404.23	106,590	746,130
Seine	5	39.36	132.51 to 171.22	197.78 to 220.74	153,764	768,819
"	6	39.36	162.36 to 213.53	131.29 to 229.93	164,205	985,228
"	1	39.36	179.42 and 94.14	124.31	300,339	300,339
Meuse[14]	1	....	150.	178.6	166,192	166,192
Ohio. [Estimated 1876.]	1	78.	400.	400 high and 400 low	463,103	....

It may be stated here that the Davis Island dam, as built, is different from these dimensions, and beginning at the right bank, has the lock, 110 feet wide, navigable pass, 559 feet, and three weirs, 224, 224, and 216 feet long, in succession. Between Davis Island and the léft bank is a fixed dam 456 feet in length.

Although the first appropriation for the Davis Island dam was made March 3d, 1875, it was not until June 21st, 1878, that the site was bought and paid for, as legislative action on the part of the State of Pennsylvania was first necessary, and subsequent legal steps taken in conformity to that act consumed more than a year's time. Work was actually begun July 24th, 1878. [15]

Under date of January 17th, 1876, a memorial from the Pittsburgh Coal Exchange, and the Steamboatmen's Association relative to the Ohio River navigation, [16] was made to the Senate and House of Representatives. This paper held that locks and dams should not be placed upon the Ohio, and that an experimental movable dam should not be built on the Ohio, as it would be both useless and injurious. The reasons for this course were given in extenso, and the cost, dangers, and consequences discussed. A system of improvement con-

---

14. 1879, II. p. 1329.   15. 1879, II, p. 1300.   16. 1877, I, p. 645.

sidered as applicable to the Ohio was outlined, and certain specific objects mentioned as worthy of appropriations. Under date of February 23d, 1877, a report[17] was made on this subject by a Board of Engineer Officers consisting of Col. Z. B. Tower, Lieut.-Col. H. G. Wright, Majors G. Weitzel and W. E. Merrill, with 1st Lieut. F. A. Mahan as recorder. The objections raised by the memorialists are answered in detail, and the fact that only a part of the navigation interests of the Ohio was connected with the opposition was dwelt upon, and it was thought that no valid reason had been shown why the proposed experimental dam should not be tried.

To meet certain of the objections raised to the proposed dam, modifications in the original plan were proposed by Colonel Merrill, and submitted to the original Board of Engineer Officers, which had approved of the first plan. This board reported[18] January 25th, 1878. "The proposition submitted by the Engineer in charge, is to give the lock a net width of 100 feet, or allowing ten feet for clearance, 110 feet, in all. This change makes it unnecessary to have an available length of more than 600 feet. The gate to close this lock will be 117 feet long, and will have a lap of three and one-half feet on each wall. The gate will somewhat resemble a railroad car, which, when out of use is housed in a recess in the bank, and when needed is run across the lock, thus shutting it up, acting as a gate. The power to move these gates and manœuvre the valves used in filling and emptying the lock chamber is to be a small steam engine, of not more than ten-horse power. We have no hesitation in affirming our belief that this gate will work satisfactorily, and that it is especially suitable to very wide locks, with low heights of wall."

After the work of construction began it was carried on continuously when the stage of water permitted work, and the funds on hand justified it. Upon July 1st, 1884, it was reported[19] the work was practically finished, except the lock gates, the machinery for moving them, and the movable dam across the head of the lock. Appropriations for this work have been made sometimes specifically and sometimes out of the general appropriations for the Ohio River.

Date of act.

March	3d,	1875.	"Towards construction of a movable dam, etc."	$100,000
"	3d,	1879.	Out of the gen'l appropriation, "not exceeding"	100,000
June	14th,	1880.	" " " "shall be expended"	100,000
March	3d,	1881.	" " " "may be expended"	150,000
"	21st,	1882.	Specific appropriation	100,000
Aug.	2d,	1882.	General appropriation, allotted therefrom	165,000
July	7th,	1887.	" " "shall be expended, or so much of it as may be necessary for completion"	70,000

$785,000

As this amount is an increase of about sixty-eight per cent. over the estimate of 1876, already given, and that estimate is about ten per cent. over the general estimate of 1875, for the whole Ohio, it

---

17. b. c. p. 637.   18. 1878, I, p. 802.   19. 1884, I, p. 260.

would appear that if this system were applied to the whole Ohio, the entire cost would be not less than $68,000,000, or about $70,000 a mile.

There is no official record, of the expression of an opinion by the officer in charge, or by a board of engineers, that it is desirable to extend the movable dam slack-water system below Davis Island; nor is it shown that such an extension would be the best solution of a plan for the radical improvement of the Ohio.

The importance of the Kanawha River, as a tributary of the Ohio, and the similarity of its characteristics to those of the upper Ohio; the intimate and similar commercial relations, and lastly the similarity in methods of improvement adopted, lead me to take up the Kanawha at this point, with the intention of resuming the Ohio later,

The Kanawha River rises in North Carolina in its principal branch, the New River, and after joining the Gauley is known as the Kanawha. The drainage area in North Carolina is about 982 square miles; and in Virginia and West Virginia about 11,922 square miles; or, 12,900 square miles in all.

Its precipitous slope from the mountain ranges, gives such rapid discharge that floods are produced unknown in other tributaries of the upper Ohio.

New River, formed by two forks, one in Ashe County and the other in Watauga County, North Carolina, has been surveyed from the mouth of Wilson Creek, four miles below this junction, to the mouth of Gauley River, where the two form the Kanawha River. Beginning at an elevation of 2,451 feet, cutting way its through the entire Appalachian chain it receives many important tributaries and drains a large area of country. At the lead mines in Wythe County, Virginia, after a fall of 333 feet, in a distance of 62 miles, it is 350 feet wide. To the mouth of Greenbrier is 128 miles with a fall of 531 feet. The Greenbrier itself is more than 60 miles long, and carries floods of 20 feet in height. From mouth of Greenbrier to the mouth of Gauley, the river passes down a wild and tremendous mountain gorge, where in a distance of 67 miles the river falls 756 feet, and flood heights of 40 to 50 feet are certain, and possible flood marks are referred to as 69 feet above low water. In this distance of 257 miles there is a fall 1,787 feet and the drainage area is so large that flood discharges of 90,000 cubic feet per second are estimated. The Gauley is another mountain stream, of over 50 miles in length from the junction of its principal tributaries. From the mouth of Gauley to the Kanawha Falls, is a distance of a mile and a half, and the entire fall is 27.6 feet. The following table shows the fall and distance from this point to the Ohio River, together with the reference above tide water.[20]

---

20. 1873, p. 837.   1876, II, p. 164.   1877, I, p. 744.

PLACES.	Dist'nce from Great Falls in miles.	Fall fr'm Great Falls basin. FEET.	Height above Sea. FEET.	Length and fall of principal shoals in low water. FEET.
Foot of Great Falls.................	0	0	618.6	
" Long Shoal..................	1.38	10.38	608.22	10.28 in 4,207
" Loup Creek Shoal..........	4.71	22.15	596.45	10.12 " 8,736
" Lykens Shoal T'n of Cannelt'n	9.19	32.10	586.50	6.19 " 2,250
Dam No. 2, located in this pool.... ..				
Foot of Harveys Shoal..	10.61	36.99	581.61	3.98 " 1,400
" Hunters Shoal............	11.28	38.09	580.51	1.60 " 950
" Windsor Shoal...............	12.39	39.50	579.10	0.83 " 1,300
" Paint Creek Shoal,..........	15.12	45.70	572.90	5.12 " 2,300
Dam No. 3, located in this pool......				
Foot of Cabin Creek Shoal...........	20.83	53.21	565.39	5.15 " 2,376
Dam No. 4, located in this pool......				
Foot of Witchers Creek Shoal.......	23.94	57.48	561.12	3.70 " 2,500
" Cat-fish Shoal head of Charleston pool—Dam No. 5............	26.49	60.29	558.31	1.59 " 1,400
City of Charleston and mouth of Elk.	36.82	60.76	557.84	Pool 10, 33 miles long, fall 0.47.
Foot of Elk Shoal.................	37.46	63.47	555.13	2.71 " 1,700
" Two Mile Shoal.............	39.33	66.55	552.05	2.79 " 1,900
" Island Shoal.................	40.05	68.85	549.75	2.21 " 1,900
Dam No. 6, in this pool.............				
Foot of Tyler Shoal..............	41.62	72.93	545.67	4.10 " 5,700
" New Comer Shoal...........	43.52	74.96	543.64	0.59 " 5,000
" Johnson Shoal..............	53.03	83.03	535.57	4.35 " 5,600
" Tacket Shoal..............	55.81	86.11	532.49	2.20 " 2,904
" Red House Shoal............	62.14	89.86	528.74	2.84 " 1,850
" Gillespies Ripple...........	67.25	93.67	524.93	0.90 " 2,200
" Knob Shoal.................	71.78	97.17	521.43	2.60 " 3,200
" Buffalo Shoal..............	73.10	98.54	520.06	0.90 " 1,000
" Five Ripples...............	76.34	102.25	516.35	2.80 "12,100
" Arbuckle Shoal..............	79.20	105.02	513.58	2.03 " 3,300
" Thirteen Mile Shoal.........	82.63	106.77	512.83	0.86 " 1,200
Point Pleasant, mouth of Kanawha...	94.20	107.92	510.68	

From this table the steepest natural low water slopes are:

Harvey's Shoals.............................. 1 foot fall in 352.
Paint Creek " .... 1 " " " 449.
Cabin " " .... 1 " " " 461.
Witchers" " .... 1 " " " 676.

All now overcome by dams 3, 4, and 5.

Elk Shoals .... 1 foot fall in 627.
Two Mile " .... 1 " " " 681.
Island " .... 1 " " " 860.

To be overcome by dam No. 6, now under construction.

Tyler Shoals .... 1 foot fall in 1390.
Johnson " .... 1 " " " 1287.
Tacket " .... 1 " " " 1320.
Red House " .... 1 " " " 651.

To illustrate the rapid change in slope, from the Falls down, the following is given, showing the slope per mile in successive, short stretches, beginning at the Falls:

Distance in Miles.	Slope per Mile. Feet.	Distance in Miles. (Continued.)	Slope per Mile. Feet.
4.71	4.70	11.41	0.88
5.90	2.53	9.11	0.75
4.51	1.92	5.11	0.74
5.71	1.33	4.53	0.77
5.86	1.21	4.56	1.10
10.65	.04	6.29	0.71
Charleston Pool		11.57	0.10
4.80	2.53		

From Cannelton, or the site of dam No. 2, to dam No. 5, at the head of the Charleston pool, 17.30 miles, the fall of 28.19 feet gives a slope of 19½ inches to the mile,— not much greater than the slope from Pittsburgh to Beaver, 26 miles. From dam No. 5, to the mouth of the river, 67.71 miles, the average slope is about 8½ inches to the mile, or nearly the same as from Beaver to Wheeling, 64 miles. The Charleston pool, and the four shoals just below Charleston, break the average, however, in a marked degree.

The discharge of the Kanawha River at extreme low water in 1881, was accurately measured at Elk chute, just below the mouth of Elk River, and found to be about 1,150 cubic feet per second through an area of 485 square feet, the gauge reading +0.17 above lowest water mark Ellet's report calls for about 1,100 cubic feet as measured in 1858. [21]Discharge at other stages found in the table.

Number of Observation.	Gauge Reading	Area, square feet.	Discharge, cubic feet per second.	Number of Observation.	Gauge Reading	Area, square feet.	Discharge, cubic feet per second.
No. 1	1.55	4,182	2,492	No. 8	10.89	9,939	28,798
No. 2	2.70	4,877	4,925	No. 9	15.55	13,169	47,120
No. 3	4.51	5,903	8,613	No. 10	18.96	15,539	58,558
No. 4	4.70	6,080	8,852	No. 11	22.11	17,710	76,851
No. 5	6.55	7,370	12,733	No. 12	26.55	20,926	98,407
No. 6	7.01	7,552	13,605	No. 13	32.85	25,365	118,291
No. 7	8.36	8,507	18,562	No. 14	34.62		155,388

These observations were taken in the Charleston pool, above the mouth of Elk. Simultaneous observations with the last gauging given, were made in Elk and a discharge of 32,959 cubic feet per

21. 1876, II. p. 160.

second determined. Below Elk then the discharge of the Kanawha will be 188,347 cubic feet per second, the gauge reading 34.'62 above low water. Floods upon September 29th, 1861, gave a gauge reading of 45.'37, and on September 12-14, 1878, showed 40.'13 by the same gauge, but no discharge measurements were made at either time. The Bellaire and Parkersburg gaugings upon the Ohio indicate that for a gauge reading of about eight feet, the Kanawha area and discharge are about sixty per cent of those of the Ohio. The rapid fall of the Kanawha and its high floods probably make its flood discharge more nearly equal that of the Ohio at its mouth, Point Pleasant.

An average of ten years gauge readings at Charleston show, [22] that

The gauge reads 3' above low water on an average 299 days in the year
And   6' "   "   "   "   "   " 148 "   "   "   "

The record being not very dissimilar from that at Pittsburgh. Prior to the beginning of work on the river by the United States, work had been done under a State charter. This work consisted in confining the water to the channel at the shoals, and in dredging, and the removal of snags, and obstructions. An appropriation of $300,000 having been made March 3d, 1875, towards the permanent improvement of the river, a board of engineer officers was convened May 5th, 1875, [23] to consider the subject. The board, consisting of Lieut. Col. H. G. Wright, and Majors W. P. Craighill and O. M. Poe, reported that movable dams should be adopted from the mouth of the river to Paint Creek, the first permanent dam to be at that point instead of at the head of Charleston pool, as suggested by Major Craighill in his project. Chanoine wickets were recommended for adoption. The depth of water to be secured was seven feet. The estimate for this plan was $4,071,216. Proposals for building the first two locks on this plan were opened July 22d, 1875. The details of the locks and dams, as built, are given in tabular form:

DAM	Class.	Inside Dimensions of Lock. feet.	Lift of Lock. feet.	Vertical height Pass Wickets. feet	Width of Pass. feet.	Length of Weir. feet.	Height of Weir Wickets feet	Begun.	Completed.	Distance below Falls. miles
No. 2	Fixed.	308x50	12					1883		9.5
No. 3	Fixed.	300x50	12					1878	1881	15.3
No. 4	Movable.	300x50	7	12'.8"	248	210	6	1875	1880	21.6
No. 5	Movable.	300x50	7	13'.	250	266	5	1875	1880	27.7
No. 6	Movable.	342x55	8¾	13'.	248	310	...	1880		40.9

Experience demonstrated that the estimated cost of each site with its lock, dam, and appurtenances was correctly placed at $350,000.[24]

22. 1882, I. p. 926.   23. 1875, II. p. 94.
24. 1880. I. p. 682.   25. 1882, I. p. 925. 1884, II. p. 929.

After completion, it is reported that the average annual expense at the movable dams, for care and operating, was about $2,555 each.[25] No great difficulties have been reported as arising during the operating of these dams, and nothing, beyond minor details, has been suggested as an improvement.

The movable dams above Charleston are in that portion of the river, which, by reason of its shoals of great fall, would not be passable, especially by tow boats going up stream with empty barges, without their aid except at reasonable stages of the water. When the stage becomes a coal tide allowing of shipments irrespective of the existence of the dams, they are lowered and loaded barges pass down the navigable pass. The influence upon business caused by the completed dams 3, 4 and 5, is shown by the record of coal passing down by the lowest lock No. 5 since completion.

	1881.	1883.	1884.
Bushels of coal passing down lock No. 5—through the pass...	875,800	4,755,103	6,217,900
Bushels of coal passing down lock No. 5—through the lock...	228.900	1,401,100	2,287,500
	1,104,700	6,156,203	8,505,400

The increase of the coal shipments by river during past years is shown:

Bushels of coal shipped by river 1875, 4,048,300; 1876, 5,024,050; 1877, 5,183,650; 1878, 5,556,050; 1881, 9,628,696; 1883, 15,370,458; 1884, 18,421,172.

But as the most of the coal is shipped from the Charleston pool, and lower, the influence of the improvements are shown by the first table only. The coal shipped from that part of the river not dependent upon the locks and dams, is, in bushels, 1881, 8,524,000; 1883, 9,214,255; 1884, 9,915,772. The increase in nine years, of more than double the production in the lower river, shows its independence of the permanent improvement, and the increase from zero to 8,505,400 bushels in the upper river, shows the value of that part of the work. The completion of lock and dam No. 2, will bring an additional coal field to the river, and meet all demands of that part of the river. The completion of lock and dam No. 6, will overcome the bad shoals below Charleston, except one, make the upward passage of empty barges at low stages much less expensive, and will furnish more water in the Charleston pool for harboring loaded boats. It will have no influence on the downward passage of loaded boats, nor will the entire system of improvement, if carried out to the mouth of the Ohio, be of service in this respect, unless the Ohio can be given the same depth, seven feet, from that point to Cincinnati. It is a matter of fact that at present, for all kinds of navigation, there is as much water in the Kanawha, below Charleston, in general, at all times, as in the Ohio below the mouth of the Kanawha.

The total amount appropriated for the improvement of the Kanawha River, up to and including the appropriation of July 5, 1884, is $1,742,000.

ELK RIVER, WEST VIRGINIA,[26]

Was surveyed from Braxton Court House to its mouth at Charleston, a distance of 100 miles. It is one of the chief tributaries of the Kanawha, draining a large area of very difficult mountainous country, rich in coal and timber; but of bad roads and difficult communication. Surveys for railroads develop such necessary cost as to deter the undertaking, at least at present. The river there is an important line to the community. Throughout the 100 miles, the low water width is about 200 feet; average fall per mile, two and one-half feet; shoals permanent, bottom rock, overlaid with stone and gravel. Frequent small rises, and occasional large ones, of twenty-five feet in height; and a low water flow of about eighty cubic feet per second are the principal characteristics. Permanent improvement is not possible without locks and dams; which can be surely and economically built and maintained. The estimate for twenty-two locks and dams; combined wood and stone structures, average length of dam 230 feet; average lift nineteen (including the five feet of navigable water) was $1,000,000, and with masonry dams $1,543,080. Appropriations were made in 1878, 1880, 1881 and 1882, amounting to $17,000, to be used for clearing the channels, chutes, and concentrating the flow. These sums have given a full equivalent in the relief furnished push boats, small flats and rafts, and have been of great advantage to a community not highly favored in their natural surroundings.

No statistics of commerce are furnished. The productions of the region are limited to logs rafted down, and supplies carried up by small craft.

MUSKINGUM RIVER, OHIO.

This river, which enters the Ohio 172 miles below Pittsburgh,[27] is the most important northern tributary above the mouth of the Wabash. It drains about 12,000 square miles, embracing in its basin nearly one-third of the State of Ohio. Its sources are within twenty miles of the shores of Lake Erie. Its basin, as a whole, has an average temperature colder than that of the Ohio Valley above the mouth of the Muskingum. Its mouth being wide and favorably situated, is much sought after as an ice harbor, at the approach of hard freezing weather. About 1,050 feet above the mouth is located the first lock and dam of a slack water improvement, built on this stream by the State of Ohio in 1838. The lock has a chamber of 130x34 feet, with a lift of 11½ feet, and is built of masonry. The pool thus formed is 5.6 miles long, and has an average width of 510 feet. The slack water extends to Zanesville, and in this distance of 65 miles, are five locks. There are small steamers plying on these pools, and necessarily limited in size by the locks. No work has been done on this stream, other than that destined to make the mouth an efficient ice harbor. This is noted elsewhere. No commercial statistics of the river are given.

---

26. 1876, II, p. 166.   27. 1879, II. p. 1366.

LITTLE KANAWHA RIVER, WEST VIRGINIA.

This stream, from near its extreme headwaters in Braxton County, has been surveyed[28] from Bulltown to its mouth at Parkersburg. In this distance of 131 miles the fall is 197½ feet. The country through which the river runs is extremely rough, with hills the entire length of it. Above the head of slack water navigation, forty-three miles from the mouth, the river bottoms are seldom more than a few hundred feet in width, and the country is very thinly settled. This part of West Virginia is covered with a fine growth of timber, principally hard wood, except along the river and its large tributaries where it has been largely cut away. The river is very narrow and very crooked, and in high water difficult to navigate with even flatboats, in descending, owing to the bends. The Little Kanawha Navigation Company, beginning work in 1867, built four locks and dams, which were completed by the end of 1874. A depth of four feet was carried by this slack-water from Parkersburg to Spring Creek, a distance of forty-three miles. All of the locks and dams were substantially built, the lock chambers being 143x23 feet, with lifts of from 10.1 to 15.7 feet, or 49.7 feet in all. Highwater mark of 1852 was thirty-eight feet above low water. From the profile any radical improvement other than by locks and dams was shown to be impossible.

By appropriations, amounting to $43,300, made in the years 1876 to 1879, all obstructions, both artificial and natural, had been removed, and subsequent appropriations, amounting to $86,000, were made for the building of a new lock and dam. The Little Kanawha Navigation Company furnish the following statistics, which cover the greater portion of the business done upon the stream:

	Rafts. No.	Coal. Bushels.	Lumber. B. M. M.	Cross Ties. M.	Staves. M.	Petroleum. bbls.
1877 [First Year.]	388	....	1,163	57	3,406	12,268
1878	436	26,450	937	69	4,978	5,084
1879	637	8,500	1,101	116	2,615	6,056
1880	908	7,890	2,437	209	2,340	3,465
1882	922	20,450	1,874	352	1,144	2,380
1883	1,208	14,900	6,230	623	1,322	4,200
1884	702	26,077	3,943	386	841	3,199

The new lock, when built, will have a lift of 12½ feet, with lock chamber 150x28 feet, and will extend the slack water navigation thirteen miles.

GUYANDOTTE RIVER, WEST VIRGINIA.

This stream rises in the Cumberland Mountains, and flowing north-westerly through a very rugged country, empties in the Ohio forty miles below the mouth of the Kanawha. Though ill-adapted to ag-

28. 1875, I. p. 743.

riculture the timber capacities of the region are great, and fine beds of bituminous coal are found for long distances on the river, and entirely unworked. From Logan Court House to the mouth of the river is a distance of eighty-one and a half miles, with a fall of one hundred and forty-seven and three-fourths feet. An old system of locks and dams, built twenty-eight years ago by the State of Virginia, fell to ruin through the decay of perishable materials. The great expense of a permanent system in a river of so great a fall, with flood heights of twenty-five to thirty feet is noted. Removal of obstructions, however, assists the downward passage of logs, and lumber in various forms, and the upward passage of the limited commerce of push boats; and the whole at small expense.

Commercial statistics for two years:

| YEAR. | RAFTS. | | | Staves. M. | Hoop Poles M. | Sawed Lumber. Feet B. M. M. |
	Poplar. cu. ft. M.	Oak. lineal ft. M.	Walnut. cu. ft. M.			
1881.	1,750	400	150	7,000	200	200
1882.	2,250	400	200	7,000	200	200

Appropriations amounting to $12,500, made in six different bills, have been of singular value to those dependent upon the unobstructed use of the river.

BIG SANDY RIVER, KENTUCKY AND WEST VIRGINIA.

[29] The Big Sandy River is formed by the junction of two forks. One, the Louisa Fork, the principal branch, rises beyond the Cumberland Mountains, in the table lands of the southwestern part of Virginia, at an elevation of 1,500 feet above tide water. Throughout its course, this stream, as well as the other fork, the Tug, flows through a rough mountainous region, which, though rich in both agricultural and mineral wealth, is very difficult of access, either by wagon road or river. The drainage area of the stream and its tributaries, equals 4,600 square miles; and, for its timber and other exports the region mainly depends upon the river. So large is the area, and steep the slopes, that rises of fifty feet occur suddenly, and yet pass away quickly, leaving in some seasons but a mere thread of water. At its mouth the low water discharge is measured as 753 cubic feet a second. From Piketon on the Louisa Fork, the virtual head of the intermittent navigation of the river, to Louisa the junction with the Tug Fork, is a distance of eighty-six and a half miles. In this distance are found forty-eight shoals, and a fall of 137'.7; or an average fall per mile of 1'.49. The average width is 200 feet. These shoals are rock bars, and have only a few inches of water on them during low stages.

---

29. 1875, I, p. 756.

The Tug Fork rises in the mountains of McDowell County, West Virginia. It has the same general features as the Louisa Fork. For eleven miles above Louisa, to the Falls of Tug, it is shallow, crooked and narrow; but above the falls the character changes, and it becomes a succession of pools separated by rock bars. The hills are very steep, and the banks alternately of rock and sand. From Warfield to Louisa is a distance of 35 miles with a fall of 61'.18, or an average 21" to the mile.

From Louisa to the Ohio River, is the Big Sandy proper. In this distance of 25 miles, is a fall of 27.5 feet; but as in low water the mouth of the river is above the low water of the Ohio, this slope will vary at different stages. There are small steamboats which ascend the river when the water permits. The average width of the river here is 300 feet. The bottom lands are about 50 feet above low water mark, and not subject to inundations except in extreme cases. The erosive action of the stream during floods is great, and the bed is covered with sand from eight to twenty feet in depth throughout this stretch.

This sand is largely carried into the Ohio, and by its constant motion down stream prevents any scheme of improvement beyond the removal of obstructions, or a lock and dam system. The first report . of survey made was dated February 24th, 1875; and Colonel Merrill reported that the only feasible way of procuring a sufficient supply of water for navigation was to canalize the river; and recommended a first lock and dam to be built near Louisa. It was also proposed to remove the large bowlders which filled the bed in places, and also to pull the snags and cut the leaning trees.

The first appropriation made was in 1878 for $12,000. Since then $192,000 more have been appropriated, of which sum $136,000 was designated for the proposed lock and dam near Louisa. The sum of $63,000 was applied to the removal of obstructions, and as a result a channel has been formed, which, at an ordinary stage of water, is fifty feet wide, and about two feet deep.

[30] This extends from Louisa about 100 miles up the Louisa Fork, and 15 miles above Piketon, Kentucky, the head of steamboat navigation. A similar channel has been formed up the Tug Fork for a distance of 108 miles. A push boat navigation has been thus much facilitated, besides the downward passage of logs. The lock to be built, is planned to have an inner measurement of 50x190 feet and a lift of 14 feet. It is not yet constructed.

As this stream is concerned mainly with the transportation of the products of a region of wholly agricultural and timber products, the exports are miscellaneous and not easy to tabulate. Reports for four years only are found.

30. 1883, II, p. 1565.

EXPORTS.	1881.	1882.	1883.	1884.
Sorghum, bbls............	2,500	8,000	6,800	6,460
Wheat, bags..............	2,000	15,000	12,750	15,937
Wool, lbs.................	115,500	170,000	144,500	158,950
Dried Peaches, lbs.......	193,500	132,000	112,200	106,590
Feathers, lbs..............	100,570	130,000	110,500	121,550
Hides, bales..............	210	700	595	655
Sawlogs and Timber, B. M., M.................	37,000	66,288	54,340	90,000
L'mb'r, sawed, B. M., M.	....	7,100	6,035	6,000
Cattle, head..............	2,000	4,200	3,570	5,355
Hogs, head...............	5,000	4,000	3,400	3,746
Chickens, number........	44,176	48,000	40,800	51,000
Estimated Value of Exports..............	$1,718,181	$1,945,366	$1,655,940	$1,832,480

CHAPTER IV.

OHIO RIVER; LOWER PORTION. ICE HARBORS, COMMERCE, AND BRIDGES.

Below the mouth of the Kanawha River the Ohio becomes more distinctly a river with sand in bed, although this is not the controlling element.

At low water the navigation is suspended, except in short pools, and for very light draft boats. Wide and shallow sand bars occur at numerous places, and between Point Pleasant and Cincinnati the task of improvement, sufficient to secure navigable depth, is yet in the future. At Twelve Pole Bar, 312 miles below Pittsburgh, 41 below the mouth of the Kanawha, a dike, built of continuous crib-work of sawed timber, filled with stone, was completed September, 1883. This dike is 2,450 feet in length, begins with the Ohio shore, and extends down stream, in a curvilinear form, far enough to contract the flow through the worst part of this very bad bar. The top is six feet above low water. Amount of material used in its construction is as follows:

Square timber, feet, board measure........................477,990
Iron bolts and spikes, pounds............................ 28,476
Rip rap stone, cubic yards,................................ 11,572
Brush, cords.. 62

The effect has not been reported as yet. Similar constructions at Four Mile Bar, above Cincinnati, have already been noted; as well as at points upon the lower river.

Similar works are estimated for other named points in the reports. The officer in charge, while repeatedly asserting that the eventual outcome will be a system of dikes throughout the river; and that possibly these may fail to give ample depth at the lowest stages; yet says undoubted temporary relief has been obtained at some places by their use; and his constructions are always made of such substantial character that the work does not need repairs. The solid foundations of these dikes enables him to rely upon their stability.

The Louisville and Portland Canal has already been briefly discussed; and the improvement of the Falls of the Ohio, themselves, so that navigation is more secure, when their passage can be effected, is a local problem in engineering, although of general commercial importance.

From tables following, it will be seen that from one-third to three-fourths of the vessels, and tonnage passing through the canal, will pass, in vessels and tonnage, over the falls; in different years; depending upon the stage of water.

Yet, a solution of the particular problem here, will not decide in determining general principles applicable to all rivers. It is well to note, however, that the detention, especially of coal fleets, and empty barges, at the Louisville and Portland Canal; owing to the importance of the busines, is so great, that additional facilities are greatly needed.

The subject of ice harbors, or protection to steamers and barges from ice, is one of great interest to their owners; and especially upon the Ohio. The increase in number of barges and flatboats, and the narrower margins of profit in the business, make those interested more watchful in such matters than formerly.

The losses of steamers, barges, and flatboats, by ice, on the Ohio, are very great. The break-ups are not always dangerous. Sometimes the ice forms and passes off without occasioning serious loss, and sometimes there may be one or more very serious break-ups during the same winter. Rivermen will often risk their boats as long as the river is open, in order to make use of good navigable water. Thus it happens in sudden cold snaps vessels are caught away from harbors of refuge, and must take shelter where they can, from the danger of the break-up, or run the risk without shelter. There are a number of natural ice harbors along the river, the most valuable of which are the mouths of those tributaries, such as the two Kanawhas, the Kentucky and the Green, which are not obstructed by bars. At these points the tributary and main river seldom break up at the same time, and boats can pass from one to another as becomes necessary. Other natural ice harbors are found below projecting points, but these generally need additional protection to give an area of suitable size. [1] The record of the necessity for such protection at Cincinnati, shows

1. 1878, I. p. 814.

that during the twenty years, from 1857 to 1877, in more than half of
the winters, there was interruption from the ice, beginning as early as
December 9th, and as late as February 23d, for periods of from four
to thirty-six days. Steamboats generally remain at their landings,
and tend to their business, as long as possible, without reference to
protection. Barges, when laid up, or in reserve, are more frequently
protected against ice, and more in need of such protection. The
disasters from the ice flood of 1876-1877 called for an inquiry from
Congress, and subsequent action. Colonel Merrill's report of the
subject is very full, and his account vivid and impressive, of the
disasters and losses at Cincinnati. When the ice broke up January
12th, 1877, after a closure beginning December 8th, 1876, there were
thirty-seven steamers in port, of which seven were sunk or carried
away; 220 barges of coal, and 407 empty coal barges, of which were
lost and carried away 71, and 175, respectively. The figures obtained
were secured by personal investigation into each loss, and were
believed to be substantially correct. The details were all fully and
exactly tabulated. Of the summary the following table is given:

CLASS.	In Port.	Destroyed or Damaged.	Loss Estimated.
Freight and Passenger Steamboats.........	15	2	$50,000
Tow-boats.....................................	8	1	350
Ferry boats...................................	4	1	3,500
Small tug and pleasure boats...............	8	5	3,700
Model barges........................	14	3	12,550
Dismantled hull............................	1	....	....
Wharf boats and ferry floats...................	12	4	3,550
United States Light House tender.........	1	....	....
Coal barges.....................................	623	248 ⎫	
Flats, etc......................................	135	37 ⎭	202,895
			$276,545

Owing to the still water in the pools of the Monongahela slack
water, heavier ice forms in this river than in the Ohio, but as this ice
has to pass over the dams, it generally reaches Pittsburgh so broken
up as to be harmless to navigation. In this exceptional break-up,
however, the high water not only carried out the Monongahela ice,
which still retained nearly its maximum strength, but it took along
with it great numbers of empty and loaded coal barges. This whole
mass went over the dams, and passed Pittsburgh on the top of the
flood. It was, of course, almost impossible to hold coal barges or
steamboats against such a pressure.

Eight steamboats were sunk, and several others much damaged,
out of the seventy wintering in the vicinity. The papers at the time
reported that one hundred and thirty-two barges, and flats of all

kinds, passed out of the Monongahela between 6:15 A. M. and 7 P. M. of January 14th. During daylight of the same day one hundred and fifty coal barges were reported as being carried along in the ice past Rochester, twenty-six miles below Pittsburgh. The total loss must have been very large, though overestimated in the papers at the time.

The bill of June 18th, 1878, gave $50,000 for "a harbor of refuge at or near Cincinnati, to protect the commerce of the Ohio River from floes of ice." This sum being too small to construct an excavated harbor, a board of engineers was convened on July 26th, 1878, to consider the question. Their report [2] has a peculiar interest, as it shows how the board tried to get opinions of various persons interested in the subject, as to the best manner and locality to answer the purpose. In all, fifty-three copies were sent to persons owning property that had been destroyed by ice, and others interested in river craft. To these, twenty-six replies more or less complete, were received. Although very difficult to analyze, and impossible to reconcile, the answers were thought to point to a conclusion already suggested by Colonel Merrill (if they could be consolidated at all), and at any rate they supported that plan as much as any other. The board adopted the same view, and it was subsequently followed out in the building of substantial dikes at Four Mile bar above Cincinnati, with the view of breaking up the ice on its way down, thus possibly rendering it less dangerous. The dikes are of timber cribs filled with stone, and rise to the level of eight-foot stage of the river. They also assist in the general improvement of the river by contracting the current and cutting a channel through a bad bar. Two dikes were built at first, and a third subsequently. Total amount appropriated, $83,000.

[3] Colonel Merrill reported on January 20th, 1879, that the mouth of the Muskingum River, between the railroad bridge and the Ohio River, which contains a harbor area of about five acres is possibly the best ice harbor between Pittsburg and Cincinnati, and it is usually full each year to its very limited capacity.

A railroad bridge elevated only forty-two feet above low water, crosses the river only 750 feet from its mouth, and 300 feet above this is located the first dam of the Muskingum River slack water improvement. This dam makes a pool more than five miles in length, with ample depth and width. The lock, communicating between this pool and the Ohio is only 130x35 feet, and is therefore too small to allow the ordinary Ohio River boats to pass up and use this pool as a harbor, supposing a draw to have been put into the bridge. The lock was also badly located. It was proposed to build a new lock of suitable dimensions, well located, and of substantial construction; and then with a proper draw-span in the railroad bridge there would be ample room for all the craft that might desire to use the harbor.

2. 1879, II, p. 1356.   3. 1879, II, p. 1365.

This work, including all minor details necessary to the plan, was estimated could be built for $225,000; assuming that the funds would be at hand when wanted. There was appropriated March 3d, 1879, $30,000; in 1880, $50,000; in 1881, $30,000; and in 1882, $40,000; and 1884, $50,000. Although construction was begun, it has not been continuous; in 1883 no funds were on hand, and the officer in charge asked $51,400, additional for completion, in 1884. The work is difficult and dangerous, the bed on which the cofferdam rests is insecure, and the presence of the dam has caused much trouble, although anticipated.

In further examinations made, with a view to ice harbors, the report for 1881 gives [4]estimates and[f]plans for building crib piers as ice breakers, at Bellaire ,and Point Pleasant, in the Kanawha River. These piers were to be 20x20 feet on top, and 25x25 feet on top and 20 feet apart; at Bellaire the estimate was $7,500, and at Point Pleasant where a double line of the larger cribs was proposed, $22,000. The [5] report for 1884, has similar examinations and plans for Middleport, Ohio, $9,500; and at Freedom, Ohio, where three cribs were proposed, and four at Elkhorn Creek, at an estimated cost of $9,000.

A joint committee of the Pittsburgh Coal Exchange, and the Steamboat Officers Association recommended, on September 13th, 1883, similar ice harbors to be made at twelve specified points, between Pittsburgh and Middleport.

There are certain features connected with ice harbors of this character which make the system difficult of application. Unless the land in the vicinity, or next to these structures belongs to the United States, vessels cannot freely use them. In the vicinity of cities, or where water front is especially valuable, the actual cost of construction must be much increased. These, and the questions of locality, to be advantageously located, make the subject more complicated, than might be thought at first. Ice breakers are now under construction in the mouth of the Kanawha, by virtue of an item of $7,500 in the clause for that river, act of 1884.

This subject is well worth more especial attention than it has received at the hands of Congress, and is deserving of more liberality. It is distinct from questions of the improvement of navigation.

COMMERCE OF THE OHIO RIVER.

Coal shipped from Pittsburgh.	Bushels.
1865	33,072,702
1866	40,000,000
1881	77,508,730
1882	81,932,730
1883	79,230,000
1884	65,930,000

4. 1881, III, p. 1951, 1955.   5. 1884, III, p. 1709, 1, 711.

STEAMBOATS LANDING, NUMBER AND TONNAGE, AND NUMBER OF
ARRIVALS AND DEPARTURES, AT CINCINNATI.

YEAR	LANDING.				SHIPMENTS.		
	Number.	Tonnage.	Arrivals.	Departures	Hog Product tons.	Ale, Beer, Whisky, Etc. barrels	Flour, Salt, Sugar and Oil. barrels.
1876............	238	....	2,779	2,808	....	....	....
1879........	227	61,782	2,725	2,730	....	....	....
1880............	234	85,280	3,163	3,172	....	....	.. ..
1881............	206	74,959	2,638	2,633	14,537	97,026	221,836
1882............	214	78,793	2,736	2,739	7,954	89,270	207,492
1883............	217	75,344	2,340	2,329	6,078	70,890	155,077

NUMBER STEAMERS REGISTERED AT PITTSBURGH.

1865...183, tonnage 22,595
1875...147, tonnage not given.

TOTAL REGISTERED ON THE OHIO RIVER IN 1875.

PORTS OF REGISTER.	Passenger.	Ferry.	Tow Boats and Freight
Pittsburgh............................	24	10	113
Wheeling..	47	23	39
Cincinnati..	57	16	27
Louisville...	27	12	20
Evansville...............	22	6	29
Total..	177	67	228

STATEMENT OF VESSELS PASSING THROUGH THE LOUISVILLE
AND PORTLAND CANAL.

Calendar Year.	PAS'NGER BOATS		TOW BOATS.		MODEL BARGES		SQUARE BARGES.	
	No.	Tonnage.	No.	Tonnage.	No	Tonnage.	No.	Tonnage.
1874 ...	343	101,016	92	14,440	.....	....	690	158,909
1875 ...	647	234,248	281	44,420	593	146,517	1,270	333,510
1876 ...	824	295,262	267	45,583	571	141,858	1,474	388,743
1877 ...	770	288,306	374	50,724	563	127,691	1,999	532,889
1878 ...	882	323,666	461	58,595	612	156,669	1,960	556,012
1879 ...	804	301,725	293	32,814	377	93,774	850	203,103
Fiscal Year.								
1880 ...	862	328,751	235	38,822	411	107,928	1,794	491,379
1881 ...	1,088	381,447	407	50,760	434	121,600	2,125	571,031
1882 ...	617	224,152	610	63,298	300	63,964	2,245	532,929
1883 ...	1,069	379,912	769	87,651	390	111,171	2,394	643,724
1884 ...	754	290,987	645	75,518	372	111,977	2,291	586,193

VESSELS PASSING UP OR DOWN FALLS OF THE OHIO.

1881 ...	272	131,322	325	40,821	59	19,493	1,067	325,725
1882 ...	376	175,328	491	68,144	128	43,682	1,538	473,332
1883 ..	256	112,658	239	37,060	73	25,275	862	285,049
1884 ...	221	102,488	242	40,020	38	12,599	1,181	382,003

TOTAL VESSELS THROUGH THE CANAL AND OVER THE FALLS.

1881 ...	1,360	512,769	732	91,581	492	141,093	3,192	896,756
1882 ...	993	399,480	1,101	131,442	428	107,646	3,783	1,005,261
1883 ...	1,325	492,570	1,008	124,711	463	136,446	3,256	928,773
1884 ...	975	393,475	887	115,538	410	124,576	3,472	968,196

BRIDGES ON THE OHIO.

While it is as necessary that railroads should cross rivers, as that
they should cross one another, it is a consequence of the passage of
rivers that a constant tax of danger, and difficulty is thereby im-
posed upon navigation interests, which the railroads do not share.
The collisions and losses at the various bridges, are more or less
frequently occurring upon all the great rivers. Upon the Ohio,
the change from single boats to large masses, controlled by a single
boat made the bridge question a more serious and difficult one to
handle, during the last twenty years. The report [6] of a board of
engineers appointed in accordance with the act of 1870, upon the then
existing bridges across the Ohio River, gave full particulars of all
of these constructions, called attention to existing dangers, and suggested remedies, and proposed the draft of a general law to cover the
subject of bridging the Ohio River.

This board called especial attention to the fact that the Newport
and Cincinnati railroad bridge, then in process of construction, under
an act of congress of 1862, modified by one of March 3d, 1869; was

6. 1871, p. 397.

a marked and dangerous obstruction to navigation. The action of the board, sustained by the Engineer Department and the Secretary of War, led finally to a modification of the law permitting bridges to be built, and an alteration of the structure itself; so that although still a serious obstacle, it does not act as it would had it been built as intended; that is, as a permanent interruption of commerce at certain stages of the river.

The services of the Engineer Department to navigation interests upon this occasion, are bearing constant fruit to this day. This board also recommended a form of a bill for the authorization of the construction of bridges across the Ohio, containing the restrictions and requirements thought necessary to protect, as far as possible, the interests of navigation; and an act of which the principal features were the same as recommended was passed and became a law December 17th, 1872. In accordance with this law when bridges are to be built across the Ohio, designs and drawings of the bridge and piers, and a map of the location, giving all necessary details for a space of one mile above and below the proposed bridge, and all necessary information, are submitted to the Secretary of War, who details a board of three engineer officers to examine and report upon the case.

Such reports and examinations are found in the annual reports whenever the occasion arises. The question of a modification of the general bridge law for the Ohio came up in congress in 1882, and the opinions of the officers experienced upon the subject were invited, and an act amendatory to the act of 1872, was passed and approved February 14th, 1883. The last act makes further provisions in protection of the interests of navigation in bridges to be constructed after its passage. Colonel Merrill reported [7] that he had attempted to get a full list of the losses by collision with the bridges upon the Ohio, but that only a partial account could be furnished, of which the following summary may be given. This report does not cover losses since June 30th, 1882.

NAME OF BRIDGE.	Date of Completion.	Interval covered when loss occurred	Steamers Struck.	Barges and Boats Injured.	Amount of loss Incurred.
Beaver...........................	1878	1878—1882	8	16	$ 28,587
Steubenville	1863	1862—1875	15	31	69,556
Bellaire..........................	1871	1867—1880	25	33	116,745
Parkersburg	1871	1868—1879	15	16	55,000
Newport and Cincinnati......	1872	1870—1882	9	12	34,625
Cincinnati Southern...........	1877	1877—1880	5	7	9,812
Louisville.......................	1871	1869—1880	10	21	70,600

7.   1882, III, p. 1926.

Between 1876 and 1882 the loss amounted to $86,073, besides the Steubenville bridge, not reported.

In a memorial sent to congress January 17th, 1876, by the Pittsburgh Coal Exchange and Steamboatmen Association it is stated: [8] "Among the serious obstacles in the way of navigation, created by the defective construction of bridges, there are two that are especially injurious, demanding the immediate interposition of congress. These are the railroad bridge at Steubenville, and the wire bridge across the Monongahela in Pittsburgh harbor. They have too long been permitted to obstruct the commerce of the Ohio, and inflict annual losses upon those engaged in carrying it on." Whilst the disposition of congress to relieve navigation of these losses, has been manifested upon the notable occasion of the rebuilding of the Rock Island bridge, in conjunction with the Rock Island Railroad Company, and in the act of 1884, where booms, dikes, piers or other structures were authorized to be built at draw openings, and raft spans to guide water craft through these openings, limiting the expense upon the part of the United States to not over $15,000 upon any one bridge in any one year; yet it would seem that if it were a matter of justice or expediency to attempt this late assistance, half measures should not be resorted to.

It would appear from the records that navigators have accepted their fate in the upper Ohio, and admitting the failure of appeal to congress, pass the Steubenville and Bellaire bridges as best they may.

There is probably no one thing that congress could do for the Ohio River of more benefit to it, than to grant a sum of money large enough to replace the narrow spans of the Steubenville bridge with one long span. The estimate made in 1871 was $200,000 for a channel span of 424 feet. A revision of the plan, and provision for a 500 foot span or longer, would probably increase the estimate, but make a permanent and valuable improvement. The next and most important change would be of the Bellaire bridge, where the channel spans are 322 and 220 feet clear openings. A bridge built at this locality now, in accordance with existing law, would be required to have a channel span of 500 feet width. The Board of Engineers in 1871 reported this bridge as securing to navigation all that the law required and as needing no change; yet the bridge was struck that year upon three consecutive days by *six different steamers*, and in 1874 by six steamers during one month.

---

8. 1877, I, p. 647.

CHAPTER V.

THE LOWER TRIBUTARIES OF THE OHIO.

KENTUCKY RIVER.

[1]The mountain torrent which is the source of the North Fork of Kentucky River, falls 1,096 feet in 4½ miles from Pound Gap, and 297 feet in the next 26 miles, and 98 feet in the next 14 miles. From this point, the mouth of Leatherwood Creek, the fall is not so precipitous, being 204 feet in 125 miles to Beattyville. The Middle Fork has a fall of 81½ feet in a distance of 36 miles, the greatest slope being 3.2 feet per mile. The South Fork has a fall of 206.7 feet in 68.5 miles. From Beattyville, 3.6 miles below the junction of the Middle Fork to Clear Creek, the head of slack water navigation is a distance of 158.7 miles, with a fall of 154 feet. The State of Kentucky built a series of five crib dams, with masonry locks, between the years 1835, and 1845, at a cost of $606,598. These dams carried slack water navigation to a point 95 miles from the Ohio; some particulars are here given:

Dam.	Distance from Ohio River. miles	Length of Pool. miles	DAMS.			LOCKS.	
			Foundation.	Height.	Length	Lift. feet.	Chamber
No. 1......	... 4.0 ...	27.0	...... Gravel	20	500	15	170′ x 38′
No. 2......	... 31.0 ...	11.0	Rock and Gravel	21	429	12	
No. 3......	... 42.0 ...	23.0	...... Rock	21	464	14	
No. 4......	... 65.0 ...	17.2	.... Rock	20	530	14½	
No. 5......	... 82.2 ...	13.0	Rock and Gravel	25	450	14½	

Having fallen into a state of decay, the dams had become useless, and general repairs were necessary; or the abandonment of the system. A survey was made of the river in 1878, and an estimate furnished, dated July 14th, 1879, of not only the repairs necessary, but also for an extension of slack-water up the Kentucky and its Forks. By the act of March 3d, 1879, an appropriation was made of $100,000 for the Kentucky, and since then, $700,000 more, of which $75,000 was specified for "a lock and movable dam" at Beattyville. The act of '84 gave $250,000; so that of the amounts prior to that time, $475,000 have been applied to the reconstruction of the locks and dams. The original estimate turned out to be much too small; and an unfortunate accident occurring at dam No. 1, in 1881, caused a large increase of expenditure. The repairs were made, however, so that navigation was resumed upon the river in March, 1872; and was at once shown to be of importance. Since then work has been continued in completing and strengthening that already done; new work being left in abeyance for the present. Lockages at lock No. 1, the lower lock, are here quoted:

---

1. 1879, II. p. 1399.

YEAR.	Steam-boats.	Barges and Flats.	Rafts.	Miscella-neous.	Lockages.
1882 (four months)...............	135	30	128	5	298
1883.................................	442	232	130	119	921
1884.................................	416	175	60	137	788

COMMERCIAL STATISTICS.

ARTICLES.	1883. (six mos.)	1884.
Iron, tons...	....	8,050
Miscellaneous Merchandise, tons.....................	1,191	5,088
Coal, bushels...	210,250	272,000
Grain, bushels	128,975	214,794
Hay, bales...	3,336	5,937
Tobacco, hhds..	2,555	4,291
Whisky, barrels..	9,233	18,216
Shingles and Staves, M........	303	2,012
Lumber, board measure, M.............................	29,664	61,437

WABASH RIVER.

This river with its branches, drains the larger portion of the State of Indiana, a part of Ohio, and quite a large part of Illinois. It is the most important northern tributary of the Ohio. Thirty-five years ago the annual value of the commerce of the river and its branches, in wheat, corn, pork and live stock alone, amounted to about $4,000,000. The worst obstruction on the river, the Grand Rapids, ninety-five miles from the Ohio, had been overcome by a lock and dam built by a chartered company. The Erie and Wabash Canal made, with the river, a through water line. The growth of railroads so completely destroyed the commerce of the stream, that the first examination made of the stream in 1871, reports that owing to the decay and destruction of the Grand Rapids lock, (built of crib work, both lock and dam) the commerce had been entirely ruined. Looking to the revival of it, the building of a new lock and dam, was the first essential, and other necessary improvements were pointed out in 1871. It so happens, however, that such navigation interests as have, since then, been developed on the river, have been mainly connected with short distances, and have been the shipment of agricultural products to a market, either by the nearest or a rival railroad.

The river flows between banks generally of a material easily cut away by the current, with consequent changes of channel. At Little Chain, Grand Chain, Warwick's Ripple and Coffee Island only, was the removal of rock necessary, and at other places the work consisted of wing dams and shore protection, cut-off dams and the removal of snags and obstructions.

The results of the complete survey are only partially given;[2] from Terre Haute down the main figures are here found:

	Distance--Miles.	Fall--Feet.
Terre Haute..	0	0
Musgraves Ripple................................	7.20	3.61
Strain's Ripple......................................	16.35	7.46
Aurora Ripple..	21.70	12.98
Bowen's Ripple......................................	25.50	15.05
Devil's Elbow Ripple.............................	35.40	23.10
Hutsonville Ferry..................................	43.50	18.17
Merom Ferry..	51.00	26.92
Greer's Ripple..	55.20	29.30
Shaw's Landing.......................................	65.70	35.03
Goose Bar..	75.15	40.98
Massey's Bend Ripple............................	85.00	46.41
Vincennes, wagon bridge.......................	90.00	48.92
Nine Mile Ripple...................................	98.85	52.28
River Deshee, head of Cat Fish Bend...	109.30	50.77
Beadle's Dam Ripple..............................	114.75	62.33
Hanging Rock...	118.75	65.18
Grand Rapids Dam.................................	121,10	69.05
Hurd's Ferry...	121.50	70.97
White River..	122.55	71.18
Coffee Island Chute...............................	128.	
New Harmony Cutoff.............................	163.	
Warwick's Ripple....................................	185.	
Grand Chain...	181.	
Little Chain...	187.	
Ohio River..	212.	

The small slope of the river, over which the levels are given, is noticeable, varying from three and a half to seven and a third inches per mile. At the Grand Rapids a fall of 4.5 feet is the most remarkable. It has been proposed to build a new lock and dam here, but as yet this is not under way. Beginning with $50,000 in 1872, and followed by the same amount in 1873, the effort was made to place the river in a navigable condition. At the Grand Chain, a channel cut through the rocks; and loose stone dikes; gave a great relief. A dam built across the cut-off at New Harmony, rock excavation at Warwick's Ripple, and a general removal of snags were the next operations. The stimulus of commerce was felt at once. From December 5th, 1877, to July 9th, 1879, there were shipped to and from Wabash Station, St. Louis and South Eastern Railroad, by boat, 14,981 tons grain; 7,052 tons flour and meal; 1,450 tons pork, beef and lard; and

---

2. 1884, III, p. 1763.

4

2,150 tons miscellaneous merchandise; total, 25,633 tons. Smaller amounts were shipped at Grayville Station, Vincennes and Cairo Railroad. Twelve steamers, given by name, took out of the Wabash River in 1879, to points near, and as far as New Orleans, 250,800 bushels grain; 580 tons pork and meat, besides other merchandise not known in amount. As the work extended over the upper river we find the shipments of agricultural products to market increase. In 1880, between Vincennes and Terre Haute, were shipped 2,158,907 bushels grain; 30,568 barrels flour; 3,790 tons general merchandise, and a large amount of lumber, staves, wood and railroad ties, while on the lower river 664,000 bushels grain, and 11,322 tons of general merchandise are reported. Appropriations were not large enough to begin the Grand Rapids lock, for which $130,000 had been estimated, but by the end of the fiscal years 1881-1882, $380,000 had been given in nine items, during ten years, to cover the distance of 212 miles; being at the rate of $1,792 per mile, or $179 per mile per annum. During 1882, 3,082,907 bushels of grain, besides large quantities of lumber and general merchandise passed over the river. As a general result of the improvement thus far made, steamers have been enabled to run and do run regularly between Terre Haute and Vincennes at all stages; and boats drawing two feet can run at all stages for more than eighty miles below Mount Carmel. Large amounts of produce and merchandise are now carried by steamboats for varying distances on the river. The bills of 1882 and 1884 gave $70,000 and $40,000 respectively, for the care of the river for the four years to 1886, being at the rate of $130 per mile per annum, or less than the average of the preceding years. A small snag boat is maintained, and minor repairs and operations carried on.

The White River, a tributary to the Wabash, has been made navigable for 23 miles, and from it has been shipped as much as 420,000 bushels of grain in one season. From 1881 to the present $50,000 have been spent on this stream. Lumber and rafts should be referred to as forming a part of the business of these rivers.

TENNESSEE RIVER.

The Tennessee River has three marked general topographical features, determined by the mountainous, rolling, and level region through which it passes.

From Knoxville to Chattanooga, a distance of 189 miles, the river is navigable during the greater part of the year. The regimen of the river is practically permanent, but little change having occurred in the past fifty years. The obstructions consisted chiefly of reefs of rock, with occasional shoals of sand and gravel. Examinations showed that by the removal of twenty-nine of these, a channel could be secured that would answer the purpose of navigation. Operations were initiated with the sum of $35,000 from the appropriation of 1871 for the Tennessee, and consisted in blasting a channel through the reefs and building stone wing dams to contract the channel, and throw as much water as practicable into it. Nearly all of the obstructions have been more or less improved; many of them are entirely removed and others reduced to secondary importance as obstructions. These rock excavations and stone dams are little affected by the elements, and there being no ice to contend with, the improvement once made is practically permanent.

The total amount appropriated for this portion of the river since 1871 is $218,500, being at the rate of $1,156 per mile, or $77 per mile per annum. As will be seen, the isolation of this part of the river from the general western river navigation, because of the Muscle Shoals, makes the commerce local to this stretch.

Statistics of river commerce at Chattanooga are incomplete, but are here quoted:

Year.	Steamers running. No.	Landings at Chattanooga. No.	Flatboats. No.	Pig Iron. tons	Iron Ore. tons	Coal. bushels	Limestone. tons	Lumber, board measure. M.	Grain bushels	Hay. bales	Miscellaneous Freight. tons
1877	10	...	...	10,000	13,700	500,000	......	8,000	1,500,000	......	......
1878	14*	...	...	......	......	......	......	......	......	......	......
1879	9	200	125	8,952	4,574	625,000	6,300	4,250	413,800	742	......
1880	9	553	308	2,540	17,677	1,001,872	10,542	6,000	592,500	......	......
1881	8†	625	396	......	......	......	......	10,440	757,564	3,540	......
1882	...	...	...	......	20,851	800,000	11,632	12,000	125,200	6,140	5,050
1883	12‡	...	...	......	......	......	......	......	......	......	......
1884	9	...	...	......	66,000	100,000	32,400	700	723,700	661	3,978

* Business not reported.   ‡ Business not reported.
† In 1881 the total amount of Pig Iron,   Only 5 of these reported their business.
  Coal, and Limestone, was 70,355 tons.

At Chattanooga a record of six years shows the river to stand
  At or above 2 feet by the gauge 325 days in the year.
  "   "   "   4 "   "   "   "   220 days in the year.
  "   "   "   6 "   "   "   "   150 days in the year.
  "   "   "   8 "   "   "   "   100 days in the year.

At low water the river is 1,200 feet wide; velocity of current two miles an hour.[3] The range between low and high water was sixty feet, in 1867. The right bank of the river is twenty-five feet above low water, the left bank sixty feet. In the vicinity of Chattanooga the Tennessee River may be considered as finally breaking through the Cumberland mountains. Unopposed in its progress so far, it forms a fair navigable stream; but when Lookout Mountain presents itself as a barrier, numerous obstructions result. Four prominent obstructions within thirteen miles, formed of sand and gravel, or of rock, are encountered; and the last, Tumbling Shoals, has a fall of four feet in a third of a mile.

3. 1868, p. 578.

Between these shoals and The Skillet, seven miles, the river flows
through a mere mountain gap, the width not exceeding 500 feet.
From Chattanooga to this point, twenty miles, the river is a succession
of basins and shoals, the total length of obstructions being two miles.
The bed of the river is of silicious limestone, of varying hardness.
The river here ends its mountain career. The ridges recede, fertile
bottom lands succeed. Reef, shoals, and eddies disappear, and the
only obstructions are sand and gravel bars. For some years after the
first allotment, in 1868, for the improvement of the Tennessee, work
was done on the points below Chattanooga, and rocks and bowlders
were removed, the most contracted places enlarged, and the worst
shoals decidedly bettered. But it was seen that until the Muscle
Shoals were made passable, such work did not produce adequate commercial advantages, and it was postponed. At Bridgeport, sixty miles
below Chattanooga, the river enters the plains of Alabama, and for
fifty-six miles has a wide level region on its right, and sand mountains
on its left. Throughout this distance the river is uniform in its
character, having a width of 1,500 feet, and excepting two reefs, an
average depth of six feet. From Bridgeport to Decatur, the river
falls perhaps twenty feet in the 115 miles. From Decatur to Brown's
Ferry, eleven miles, is an average depth of nine feet; least depth
three and a half feet; width 1,500 to 1,800 feet.

[4] Brown's Ferry is the head of the lower group of shoals. Of these
the upper, or Elk River Shoals, has a fall of 27 feet in 12½ miles,
with a maximum fall of 3'.4 in 8,300 feet.

From Lamb's Ferry, one-fifth of a mile below the foot of Elk
Shoals, to Bainbridge Ferry, lies a series of cascades, and shoals called
the Muscle Shoals.· In this distance of 17.7 miles the fall is 84.6 feet.
This barrier to navigation is absolute and formidable. Between Big
and Little Muscle Shoals are three miles of good water, and the last
named shoals are three miles in length. In this distance of 6.3 miles,
from Bainbridge ferry to the railroad crossing at Florence, the fall is
22.6 feet. Altogether, from Brown's Ferry to Florence, in a distance
of 38.5 miles, is a total fall of 134 feet. This tremendous natural
barrier absolutely cut in two, for purposes of navigation, the entire
Tennessee system. For, 367 miles of navigable water above Muscle
Shoals, besides hundreds of miles which could be made navigable, on
the Tennessee, and its tributaries, (estimated in 1868 to amount, in
all, to nearly 1,300 miles), form a system independent and isolated;
and below Florence 255 miles to Paducah, and there connecting with
the Ohio River, make the lower portion. As early as 1824 it was
proposed to overcome this barrier by a canal, and in 1828 Congress
allotted 400,000 acres of land to the State of Alabama as an assistance towards building it. From 1832-1837 work was carried on and
a portion of the canal built. It was never completed, however, and
the portion done was of no use, and fell into decay.

4. 1877, I. p. 584.

In 1872 an instrumental survey of the Muscle Shoals was made, upon which the present project of improvement was based, but modified in 1877.

Below Florence the principal obstruction is Colbert Shoals, eleven miles below Florence, 236 below Chattanooga. Rock reefs of tough flint, limestone, bowlders, and large rocks, besides a sharp fall, and slight depth of water, composed this obstruction. In 2.44 miles the fall is 10.65 feet. Below Colbert Shoals is a pool of 3.6 miles; and then Bee Tree Shoals, marked by large bowlders, and a fall of 5.59 feet in 1.6 miles, are the last serious obstacles to navigation on the lower river. At Bear Creek Shoals, 261 miles, Big Bend Shoals, 281 miles below Chattanooga, are gravel bars or large bowlders; and sand bars at other points lower down. Duck Creek Shoals, and one or two other points, received some special work, but in general little has been found necessary below Colbert Shoals.

The great work of the Muscle Shoals Canal has absorbed the greater part of the funds possible to be so used since work on the Tennessee was undertaken.

[5]Surveyed and projected in 1868; afterwards more minutely in 1872, and with careful estimates; it was not until 1875 that a sum, sufficient to inaugurate the work with safety, was appropriated. Minute surveys were completed in 1874; estimates furnished in the spring of 1875; and contracts made that fall and winter. Subsequently the project then formed was modified, to advantage as to the Elk River Division, in 1877, and a less expensive plan adopted.

A series of delays occurred in connection with the contracts so that satisfactory progress was not made before the middle of 1879. Since that time work has progressed rapidly whenever funds allowed, but a delay of several years has arisen from lack of sufficient appropriation.

The existing project consists in enlarging and rebuilding the old canal 14½ miles long around Muscle Shoals, together with 1½ mile of new canal on the Elk River division; and of new channels blasted in the solid rock, removal of bowlders and obstructions, and the building of stone wing dams at Elk River and Little Muscle Shoals, to confine the water and deepen the channel. For this the original estimates were $4,133,000. In the 36 miles embraced in the project 8 miles need no work; 16 miles are overcome by canal and 12 miles have been improved by work in the river bed. This work, done by means of temporary dams, diverting the river from the part of the channel under operation or by the use of coffer dams, involved the removal of about 100,000 cubic yards of rock bed, in 2½ miles of the channel. The improved channel is smooth and uniform in section, 110 to 120 feet wide and 3 feet deep at extreme low water. The permanent stone dams aggregate about 3 miles in length and contain

5. 1884, II, p. 1641.

over 80,000 cubic yards of stone. The upper canal will have two locks of which one has 12-foot lift, and the other 5 to 10 feet, depending on the stage of water. The foundation pit of one has been excavated and about one-seventh of the canal trunk finished. Of the lower canal five-sixths has been enlarged, deepened and straightened by excavating 634,000 cubic yards of earth and 88,000 cubic yards of solid rock. The upper end has been extended into the river and an embankment of the excavated rock, 800 feet in length, makes a safe upper harbor. Nine new locks have an aggregate lift of 84 feet, and an extreme lift of 94 feet. These locks have chambers 60 feet wide and 300 feet between gates. The canal trunk will be 70 feet and 120 wide, and a navigable depth of 5 feet will be found throughout. The entire masonry of the nine locks is completed and they are ready for the gates. The canal will cross Shoal Creek, a tributary draining an area of 800 square miles, by means of an aqueduct of iron or wooden trunk resting upon 25 piers and 2 abutments. These are of masonry, each 75 feet long and average 11 feet in height, and are all built. All the masonry is founded upon solid rock, and in many cases the bed rock has been blasted to a greater or less amount.

At Bluewater Creek is a permanent bridge used in construction, having seven piers and two abutments, which are grooved at the sides to receive the sections of a permanent dam when the canal is ready for filling.

At Second Creek is a similar construction, and a stone waste weir at Douglas Branch.

No work of construction was done after June, 1883, there being no funds available, until after the passage of the act July 5th, 1884, which gave $350,000. The original estimate for the Tennessee River below Chattanooga was $4,133,000. There has been allotted and appropriated $2,695,500. There is hardly any doubt that if the amount asked for by the Engineer Officer in charge were given him at once, this canal could be completed within a year and thereon open to commerce an almost continuously available line of water intercourse for local and general traffic in 622 miles, passing through parts of three States—not to allude to the interest of the State of Georgia or to the Ohio River interests.

Although the commerce of the Tennessee has been thus hopelessly hampered, it is not entirely gone. Steamers ascend to Florence, and meagre statistics are given of their business. In 1881 three lines of steamboats, in 1882 five, besides five towboats, and in 1884 eight steamboats were employed in carrying cotton, 25,000 to 30,000 bales; grain, lumber; pig iron, 7,000 to 10,000 tons; peanuts, tobacco, lime, salt, flour, and amounting in all to between 25,000 and 50,000 tons of miscellaneous merchandise per annum. The coal, iron ore, timber and lumber business of the upper river being separated from the lower, it is impossible to foretell what influence the opening of the canal will have upon trade, but it is doubtless a question of some years before commerce could readjust itself to the new conditions.

Railroad bridges at Florence and Johnsonville must be furnished with ample and well located draws before navigation can be easy, and the special recommendations of the officer in charge, Maj. W. R. King, on this subject invite congressional action.

Of the tributaries of the Tennessee, the French Broad from Leadvale to Knoxville, a distance of 90 miles, is navigable now by flat boats, but it thus furnishes a commerce of a certain value. In 1882, 184 keel and flat boats brought down 86,200 bushels of grain, 1,200,000 feet lumber, and 179 tons miscellaneous freight; and 120 returned with 581 tons. In 1884 large quantities of lumber, grain and tan bark were boated down to Knoxville. The improvements already made are permanent and useful.

Amount appropriated 1880-4, $22,500.

HIWASSEE RIVER.

This stream has been worked upon from the mouth to Savannah Ford, 33 miles. Commerce is carried on partly by steamboats from the Tennessee, partly by flat and keel boats. Improvement valuable and practically permanent. Commerce in 1883, 66,336 bushels grain, 743 bales cotton, 5,000 bushels cotton seed, besides lumber and rafts.

Total appropriated 1876-1884, $31,500.

CLINCH RIVER.

The drainage area of this stream is 1,436 square miles; its length is 230 miles; and operations have been carried on to secure 2' at low water from the mouth to Clinton, 70 miles, and 1½' to Haynes, 75 miles further. Improvements made at the worst points on this river have been of marked value; in removing obstructions and deepening the channel.

Commerce is done principally by flat bottomed boats, and rafts, but occasionally steamers are used. In 1881-2-3 it is reported as follows:

```
Rafts 2,943, of 197,726 logs, board measure, feet......49,181,500
Flatboats 350, black walnut, board measure, feet....... 5,300,000
   "       "    zinc ore, tons...............................  2,000
   "       "    grain, bushels.............................  200,000
   "       "    potatoes, bushels..........................   53,000
Steamers 5, flour and potatoes, barrels....................   2,550
   "        grain, bushels.................................  110,000
   "        bacon, tons.......................................   400
Towboats 3, and 13 barges, took coal, bushels.........   70,000
   "        "        "      "   iron ore, tons.........   35,000
```

In 1884, 250 flatboats passed Clinton with about 600,000 bushels grain; and also 1,000 rafts passed. But one steamer made a trip on the river.

Total amount appropriated 1880-1884, $21,000.

DUCK RIVER

Is now in fairly navigable condition from Centreville to the mouth, 68 miles, work having been done on twenty-one shoals. Three small steamers were on the river in 1884; making 68 trips, carrying corn and peanuts. There were 25 rafts of 500,000 feet B. M. In 1883, two steamers carried 50,000 sacks grain.

Total amount appropriated 1880-1883, $13,000.

Upon the Little Tennessee, one appropriation of $5,000 was used in the usual way, and 46 rafts and 150 flatboats are reported in 1884. Small steamboats occasionally ascend this stream.

In this system we find five tributaries of the Tennessee of which 348 miles, in all, have been interested by the expenditure of $93,000 during ten years. These streams are evidently all used more or less as avenues of commerce for timber and farm products, and yet the amount allotted them has not exceeded $260 a mile. The areas concerned are in the eastern and southern parts of Tennessee.

CUMBERLAND RIVER.

The Cumberland River rises on the west slope of the Cumberland mountains. With a rapid descent, it reaches the first bench of the highlands, and after a precipitous fall, it flows with a more equable current through the highland plateau. Cutting a pathway through solid rock it reaches the Great Falls, where the total fall is 63 feet.

From the Falls to the mouth of Laurel River, 10 miles, the river flows between cliffs, which rise, occasionally, 300 or 400 feet; and the descent is 85 feet. From Laurel River to Smith's Shoals, 24 miles, the fall is 31 feet. The elevation of the bluffs is not so great here, but the country near the river is broken by ridges and ravines.

Smith's Shoals are rapids, covering a stretch of nine miles, and having a fall of 54 feet in all. The shoals are formed by ledges of limestone. Throughout their length the bluffs are 300 to 400 feet in height. The pool below the shoals is two miles in length, and has a fall of seven-tenths of a foot. Point Burnside, at this point, is the head of navigation.

⁰ To Nashville the following table shows distances and slopes :

	Distance from Cincinnati Southern R. R Bridge miles	Fall from initial point. feet	Slope per mile feet	Least depth found at low water on shoals feet
Gann's Ripple..................................	29.58	16.704	0.591	0.4
Blankenship Island..........................	59.44	43.854	0.909	0.9
Crocus Creek Shoal.........................	80.60	57.913	0.664	0.7
Bear Creek Island...........................	91.46	61.023	0.286	1.0
Stalcup's Island.............	108.75	77.580	0.958	1.0
Weaver's Bar...................................	141.48	96.671	0.583	1.0
Scantlin Island..............	150.67	105.887	1.001	0.9
Gainsborough Pool.......	161.47	108.183	0.217	1.2
McCarver Island..............................	171.80	121.197	1.260	0.7
Upper Holliman Island.....................	181.56	125.062	0.396	0.7
Sullivan Island...............................	188.85	136.081	1.511	0.7
Lovell's Island...............................	220.08	153.686	0.563	0.7
Hartsville Island............................	239.78	171.054	0.929	0.9
Station Camp Shoal.........................	280.30	196.652	0.629	0.5
Lower Nashville Island.....................	327.83	222.955	0.553	0.7

This table has been arranged so as to bring out the differences of slope; and yet it will be noted how uniform, in general, the slope is. In this stretch the banks are generally high enough to prevent overflow, and the river bed is commonly rock. The range between high and low water at Point Burnside is 65.5 feet; at Nashville, 52.9. To Point Burnside the river is navigable from four to six months in the year, for steamers of three feet draught and less; to Burksville, 238 miles from Nashville, the season is from five to seven months; and to Carthage, 118 miles, from six to eight months.

Below Nashville to the mouth of the river the distance is 192 miles; the fall 79 feet, and the slope about 0.41 feet per mile. The obstructions known as Harpeth Shoals are the most serious. The length is 4.3 miles; fall 11.59 feet; and maximum slope 7.81 feet, near the lower part of the island. Between these and Davis' Ripple the alluvial bottoms increase in extent, and no serious obstruction is found until the last named place. Here an average slope of 3.77 feet per mile is found, over a distance of 1.34 miles, and a maximum slope of 9.16 feet per mile for a short distance.

From Davis Ripple to Clarksville, 65 miles below Nashville, the river winds in long curves. Palmyra Island, 75 miles below Nashville, Elk Creek, Dover Creek, Middle and Lower Gatlin, Race Track, Little River and Ingram Shoals are minor obstructions. The last one, 148 miles below Nashville, with a maximum slope of 7.10 feet per mile, was the last serious obstruction and called for considerable work in dredging and the building of wing dams. In successive river and harbor acts the Cumberland has been divided into a number of different sections, but as the obstacles to be overcome are of a similar nature in these, the method of improvement was the same, and consisted in blasting out a channel through the rock reef, removing gravel bars and bowlders, building rip rap dams and removing snags and overhanging trees. The work done has had the effect of extending the duration of navigation both above and below Nashville. In 1884 small boats ran on the lower Cumberland during the entire year. During the winter of 1883 and until May 1st, steamers ran up as far as Burnside. Statistics of commerce on the Cumberland are not full.

YEAR.	Number Steamers.	Grain. bushels.	Tobacco. hhds.	Flour and Salt. bbls.	Miscellane's Merchandise. tons.	Passengers.
1877	14	200,760	5,487	17,823	10,000	....
1884	20	1,916,870	4,919	13,735	10,395	15,215

In 1880, 16 steamers; in 1881, 9, from 300 to 500 tons, and 5 from 125 to 260 tons, were engaged in trade.

Flat boating is not carried on extensively except to bring out coal from points above Point Burnside, and this is only practicable during occasional high stages, about 4 or 5 during the year; 34 barges in 1884. Some produce and tobacco came to market in small flat boats.

Appropriated above mouth of Jellico, Ky.............$ 15,000
" Smiths Shoals............................ 115,000
" Nashville to foot of Smiths Shoals.... 162,000
" below Nashville.......................... 242,500

In the 528 miles of navigable waters upon the Cumberland River the average appropriated per mile is $983. During the 15 years, 1871 to 1886, the average per mile per annum is $67.70.

Of the tributaries of the Cumberland, three have received small appropriations. Obey's River enters the Tennessee about 3 miles south of the Kentucky line. The head of navigation is at Barnes Landing, 43 miles up. Small boats have ascended to Eastport, 18 miles higher. In 1882, 60 rafts, and in 1883, 300 rafts are reported, but no steamboats.

Total appropriated, 1880-1882, $11,500.

Caney Fork River, rising in the Cumberland Mountains, is navigable from Sligo Landing to its mouth, near Carthage, 120 miles above Nashville. In this distance of 80 miles the fall is 50 feet. Certain bad bars and rock obstructions have been removed to the great advantage of navigation. Steamer from the Cumberland made 15 trips in 1883, and 24 trips in 1884, carrying miscellaneous merchandise and in increased quantity.

Total appropriated, 1880-1884, $17,000.

An appropriation of $5,000 made in 1881 for the Red River, a small stream, was used over 38 miles, and materially assisted the commerce of that distance.

A careful review of the Cumberland and Tennessee systems of navigable waters, rising in regions rich in timber, coal, and ores of iron and zinc; flowing through a difficult terrain, where railroads and common roads cannot be extensively sub-divided, and where a sparsely settled area naturally is benefitted by open navigable streams which can float its heavy products to lower markets—a careful review of this situation must show the expediency of a full development of the possibilities of navigation, even if confined to rafts, push boats and small flat boats The permanence and simplicity of the needed improvements make the work satisfactory, and for all, except the Muscle Shoals, the results are fully commensurate with the expenditures. Had funds for the great canal been on hand and the work finished as it could have been, years ago, the whole system must have been of decided benefit to the whole of Tennessee, southeastern and southwestern Kentucky, northern Alabama, and a part of Georgia.

CHAPTER VI.

MISSISSIPPI, MISSOURI, AND ARKANSAS RIVERS.

REMOVALS OF SNAGS, WRECKS, AND OBSTRUCTIONS.

The bill of June 23d, 1866, contained an item of $550,000 for the "improvement of the Mississippi, Missouri, Ohio, and Arkansas Rivers;" and a second item of $550,000 for "construction of snag boats and other apparatus for clearing western rivers, and for the outfit, working and preservation thereof." An allotment of $172,000 was made to the Ohio River, out of the first item, and Col. J. N. Macomb was placed in charge of the general work indicated by the bills.

The close of the war found these rivers encumbered with the wrecks of many years, besides the accumulations of natural obstructions, due to an interval of twenty years. The construction of boats best suited to remove these, was a work of almost original design; and the operating of the boats, when built, a trade yet to be learned. Within six months, models and specifications were prepared for a snag boat, in which the principal features were: machinery for hauling out snags without interference with the propelling machinery; auxiliary engines for various duties; and strength of hull. Proposals were at once invited for building hulls, and contracts made in May, 1867, for three snag boats complete. Meantime, work was done by a steamer purchased and fitted up for the purpose, and a contract was made for special work upon the Arkansas and Missouri. The contracts for the three new boats amounted to about $192,300, or say $64,000 each; and the snagging upon the Missouri was contracted for at the rate of $275 per day; on the Arkansas at $160 per day. Of the new boats the first, the J. J. Abert, began work March 28th, 1868, the S. H. Long, April 25th, 1868, and the R. E. DeRussey, May 11th, 1868. Subsequently a snag boat of lighter draft was built, for use principally on the Arkansas River. This boat, the S. Thayer, began work in May, 1869. The contract price was $27,500. A dredge boat, the Octavia, was chartered in 1868, and subsequently bought, and used for dredging and scraping channels over bad bars in time of low water; besides assisting the snag boats in many ways. The fleet of boats thus constructed operated in the Mississippi, Missouri, and Arkansas Rivers; occasionally doing work upon other streams. Work upon the Ohio was begun by these boats, but subsequently carried on separately, under the officer in charge of that stream; and all appropriations beginning with the second, or that of July 11th, 1870, under the title of "Improvement of the Mississippi, Missouri, and Arkansas Rivers," were applied to the continuance of this work. The boats, as built, were maintained, by suitable repairs, in operation until 1873; when two, the Thayer and Abert, were so worn out as to be, of necessity, withdrawn from active service. Up to this time the work done is summarized as follows:

CONDENSED STATEMENT OF SNAGGING OPERATIONS FROM
MARCH 28, 1868, TO JUNE 30, 1873.

RIVERS.	NAMES OF BOATS.	Snags pulled.	Trees cut.	Drift piles remov'd
Missouri....	R. E. DeRussy, S. Thayer, J. J. Abert,	8,119	40,846	212
Mississippi	ditto, [S. H. Long, Octavia.	6,001	41,772	47
White	R. E. DeRussy, S. Thayer...............	778	422	3
Arkansas..	S. H. Long, S. Thayer...................	1,930	2,714	8
Ouachita ..	Octavia, R. E. DeRussy.................	499	24,398	7
St. Francis	J. J. Abert, S. Thayer..................	431	151	2
	Total.......................	17,758	110,303	279

1873, p. 491.

From the two appropriations of 1866 there was available...$ 928,000
The acts of 1870 and 1871 gave each $150,000............... 300,000
And the act of 1872... 90,000
  Or to cover the period of operations from 1866 to   ⸻
    1873, seven years.....................................$1,318,000

When Colonel Macomb was relieved from the charge of this work in July, 1870, he speaks of the great efficiency of the fleet and the experience learned. Lieut.-Col. Raynolds, his successor, says that as one direct result boats were able to run at night upon the Missouri River, a practice previously impossible; and that thereby a saving of four day's time in a round trip from St. Louis to St. Joseph must have been effected.

Much of the efficiency of the fleet, and the design of the old boats, and all of the details of construction of the new iron-hull boats, afterwards built to replace them; are due to Captain, now Major C. R. Suter, the officer directly in charge of the boats, and after April 7th, 1873, in entire charge of the whole, relieving Col. J. H. Simpson at that time.

A beginning was made February 5th, 1873, towards replacing the old fleet with one of permanent construction, by a contract then made for an iron hull, intended to take the Abert's machinery. The contract price was $112,500. After much and unexpected delay this new boat, the J. N. Macomb, began work December 10th, 1874. In 1878 contracts were let for two new boats, both iron-hulls, to take the machinery of the Thayer and the Long. One, the C. B. Reese, was finished in December, 1879, and set to work on the Arkansas. This boat is a stern-wheel, is 175 feet long, 36 feet wide, draws light, 28 inches aft and 20 inches at the bow, which, with the deck, is suitably shaped for her work.

The boat has 4 steam capstans, a pair of heavy iron shears and a sweep chain, and uses a powerful force pump, besides the other ordinary machinery usually used. The second iron boat, the H. G. Wright, was finished December 23, 1880, is 187 feet long and 62 feet wide, drawing 30 inches when in working trim. This boat is quite similar to the Macomb in the arrangement of hull and machinery, but draws 12 inches less, and weighs with crew, stores and fuel on board and fully equipped for work, 750 tons.

[1] Experience with the Macomb showed the value of the iron hull. Six years' service was had without any injury and with scarcely the loss of a days' service from that cause.

The average cost of repairs to the hull, which repairs consisted only in paint to prevent oxidization, was less than $500 per annum, and at the end of that time the boat was as good as new. The wooden hulls, which this succeeded, cost annually $5,000 and $6,000 to keep in repair, with an additional charge at the end of five years of nearly the first cost in order to make them last ten year, when reconstruction became necessary.

Building and maintaining a wooden hull boat for ten years amounts nearly to $2\frac{1}{2}$ times the first cost, while at the end of that time all but the machinery needs replacing.

The iron boat for the same time costs about the same, but at the end of the time is still good for many years of efficient service. Besides this iron boats are of lighter draft and do not become water logged as do wooden hulls. Their great structural strength enables them to endure shocks which would wreck a wooden boat.

1. 1881, II, p. 1600.

In 1882 plans and working drawings were prepared for a new iron-hull boat for work on the Missouri, but the condition of the funds available have not allowed of the building of this boat before 1884. The operations upon the Arkansas as well as the kindred rivers, White and St. Francis, were transferred to the officer in charge of that district in 1881. A general summary of the work done during the past eleven years will be found below, which, with the prior table, will show the magnitude and importance of the work. In fact, as long as the interests of navigation upon these three rivers are considered of national importance, it must be acknowledged that this work should be continued. It is also plain to be seen that the amount needed is about the same annually, and probably will continue so to be, and now that a permanent and efficient fleet is almost completed, as recommended by the officers in charge, no better investment could be made than the constant regular employment of these boats during the suitable seasons of the year.

SNAGGING OPERATIONS ON THE MISSISSIPPI, MISSOURI, AND ARKANSAS RIVERS, 1874-1884. ANNUAL TOTALS.

YEARS.	Appropriations	Snags pulled. No.	Weight of same Tons	Trees cut. No.	Drift piles removed. No.	Miles run. No.	Boats in operation. No.
1874	$ 100,000	1,471	20,487	1,923	20	6,742	3
1875	100,000	2,773	63,582	5,658	46	7,365	3
1876	100,000	2,214	38,825	3,206	18	5,782	2
1877	50,000	1,030	17,444	637	12	3,106	2
1878	40,000	1,712	27,903	4,523	24	5,488	2
1879	170,000	2,009	26,923	2,948	35	4,204	2
1880	190,000	3,064	53,168	1,352	44	7,391	3
1881	200,000	2,651	36,767	4,128	52	6,466	4
1882	220,000	4,601	62,747	12,930	44	8,380	5
1883	185,000	5,324	88,166	31,512	54	10,767	5
1884		2,251	27,564	1,148	23	2,634	4

Miles run on Arkansas in 1883 not given.

Miles run on Arkansas, and weight of snags pulled for 1884, not given.

CHAPTER VII.

MINOR RIVERS OBSTRUCTED BY SNAGS, LOGS, WRECKS AND PERMANENT OBSTRUCTIONS.

There is a class of many, and in some cases important streams, which are characterised as above. On some the obstructions are recurring and need constant work, and on others when thoroughly removed once they demand but little additional care. It will answer our purpose best to select certain examples and give some details, and by tabulating some others a general view may be given which will enable certain deductions to be drawn.

It will be seen that in general these rivers are found where in a densely timbered country, and with little slope, the stream meanders with a sluggish current at low stages; and as such conditions are nearly always in the alluvial regions, these rivers will be found in the southern and western regions from North Carolina to Arkansas.

RED RIVER OF LOUISANA.

The size, length and commercial importance of this river warrant more attention than the intent of this volume will justify. But as an example of thorough obstruction from logs, snags, and drift piles and incident dangers, nothing could be more marked than the so-called Red River raft, lying along and in a course of the river near the boundary line of Arkansas. Before its removal was finally attempted the head of the raft was about 75 miles above Shreveport, and about 190 miles below Fulton, Ark. Shreveport is 330 miles above the mouth of Red River, and was the practical head of navigation. The raft, although it could be passed at times, debarred the upper river from the privileges of navigation. Besides its injurious effects in this way, the alteration of the free flow of floods and the submergence of valuable land made it of baleful importance. Of historical interest and note, it was the subject of theory as to cause, effect and removal. An attempt to remove it during 1841-44 resulted in failure. A survey of the region made in 1871 and 1872 gives a description which [1] will be quoted and an initial appropriation of $150,000 was made June 10th, 1872, for the removal of the raft. The country in the immediate vicinity of the raft is one of densely wooded alluvial bottoms, and the upper portion of the river with much of the territory of many of its tributaries is similarly characterised. Near the rafts the valley varies from three to eight miles in width, and between the limits of the bluffs winds in a slowly changing bed.

The red alluvial soil covers all of this valley and on it grow cotton wood, sycamore, black and sweet gum, hackberry, ash, oak, cypress and willow in profusion. After freshets, caving banks throw im-

---

1. 1878, p. 635.

mense quantities of trees into the stream, which are stranded on bars
and shoals, and when drier are carried by subsequent floods down
until stopped again. Constant and successive damming of the stream
by these logs leads to the formation of outlets and the shifting of the
river. Mud rapidly covers logs lying in the bottom, and below the
surface near these jams. Willows spring up on decayed logs, and
islands and new banks are as rapidly formed, almost, as eroded. The
raft which obstructed the river in 1872, from Carolina Bluffs to within
four and a half miles (by river), from the Arkansas line corresponded
with the condition as described in 1805, though attempts had been
made to remove it as late as 1844. The presence of the raft tended
by its obstruction to gradually increase the elevation of the immediate banks and its bayous, and this was said to amount to about three
feet in thirty years preceding. After sixteen years the head of the
raft was found thirteen and a half miles, by the river, higher than in
1854. The total length of the channel to be opened was seven miles.
The whole area of floating raft amounted to 290 acres. The whole
area of tow heads, or rafts resting on the bottom and above water, at
the time of the survey, was 103 acres. The opening of the channel
did not necessitate the removal of all this. Navigation found insecure
and circuitous routes around the obstructions through Red and Black
Bayous at low water, and Kelley Bayou or Posten Bayou in high
water.

REMOVAL OF THE RED RIVER RAFT.

The steamer Aid was purchased in Pittsburgh[2] for this work. This
boat was built with two hulls, each 136 feet long, 15 feet beam,
double ends, and separated by a space of 14 feet. The wheel was
placed near the stern between the hulls and protected by them.
Engines were 15 inches diameter, $4\frac{1}{2}$ feet stroke; boilers three in
number, 24 feet long, 36 inches diameter, two flues each, and carried
136 pounds steam. The outfit consisted of one steam capstan on the
starboard bow, a hand capstan on each stern, three heavy double
geared hand crabs, besides two light ones, and a railway hoisting
carriage with four double geared windlasses. The boat was thoroughly
overhauled and strengthened, and there were added : an additional
steam capstan on the bow, two large boom cranes, a sloping apron at
the bow between the hulls, a steam force pump of high power, four
portable steam saws, and two portable boilers; and the boat was
assisted by two crane boats, flat boats for quarters, and machinery for
minor work. The Aid reached Shreveport January 10th, 1873.
Shore parties had already begun the work of removing such logs as
could be moved, chopping leaning trees and girdling others. The
Aid began work January 27th, pulling snags, and blasting a channel
through the masses of logs, sawing the larger logs into smaller and
more easily handled fragments, and by the end of January the raft
was attacked at five different points by various parties. By the 16th

2. 1873, p. 613.

of May a narrow channel had been opened through the twenty-sixth and last raft, and that day a boat passed through, laden with freight for the upper river, the first boat in twenty-nine years so passing. About 10,000 bales of cotton passed out from the upper river that season.

In opening a canal for the passage of drift from one of the upper rafts just above it, into Posten Lake, use was first successfully made of nitro-glycerine in the removal of growing trees in deep water, and thereafter its use was of great service. Trees with a hundred feet of trunk, and a full top of branches above the water, were blown down by charges placed beneath the water, and were cut entirely across. Masses of heavy logs were easily moved, refractory snags and large irregular stumps were blown to fragments. The remainder of the season was employed in widening and bettering the narrow channel opened; and the work of the six months consisted in the opening of a navigable channel through four miles of raft, by which the navigation from Shreveport to the upper Red River, was very much improved, and the time of possible navigation increased by avoiding the route through Black Bayou. Three and a half miles of raft remained to be opened. [3] When the water permitted it, work was resumed, and by the 27th of November, 1873, the remaining portion of the raft obstructing the channel went out, and until the end of the calendar year work was continued in widening and clearing out fresh deposits.

After this time a steamer with crane and shear boats, was employed, when funds permitted and necessity called for, in breaking up the jams which frequently reformed, and removing all snags and drifts. In 1874 and 1875, this work was quite extensive. In 1876,[4] the total length of forty-two rafts was given as about two and one-fourth miles, with a width of channel through the whole of not less than one hundred feet. Later in the season high water and much drift effectually closed the channel for two months, and for the remainder of the year entailed much work.

In 1879 a sudden rise carrying away two spans of the railroad bridge at Fulton, Arkansas, and bringing down a very large amount of drift, entirely closed the channel by jams, having a total length of five miles, which were removed as soon as possible by the steamer Florence. Since the opening of the raft the caving of banks had been much lessened, and the river had scoured greatly, lowering the high water level, thus reclaiming for cultivation much valuable land. These operations were afterwards extended in 1880 by beginning at Fulton and cutting a channel in a similar way through the many and various obstructions found in the 130 miles below that point. Since that time constant patrolling of the river by the U. S. steamer Flor-

---

3. 1874, I, p. 702.   4. 1876, I, p. 596.

4

ence has been found necessary to keep the river fairly open, and there is no prospect that there can safely be an intermission. Over the lower river it is found necessary to operate constantly with a snag boat, equipped for handling any obstruction met. The last boat built for this purpose has an iron hull. This boat, the C. W. Howell, began operations September, 1882, and finds constant employment during the working season.

The closing of Tones Bayou designed to concentrate the waters of Red River in a bad stretch of the river need not be discussed in this place although included by name in the appropriations for the removal of the raft.

Appropriations as follows:

Removing raft in Red River and closing Tones Bayou, La...$419,500
Removing obstructions from Red River, La...................... 117,500
Improv'g upper Red River from Fulton, Ark., to head of raft. 20,000
Whole of Red River.. 130,000

Total... ....$687,000

The question of the commerce of the upper Red River is not easily discussed, and that the whole river is itself somewhat indefinite. It is stated [5] that the cotton crop of 1870 was got through the raft at great expense, and much damaged by handling. Four small boats succeeded in getting through the raft, by way of a slough, and brought the cotton, 22,000 bales, to Bargetown, whence it was carried over from the river to Moon Lake, and there put on flat boats to be taken some distance farther to meet the steamers from below coming as high as possible. After the opening of the raft the amount of cotton reaching Shreveport from above is not always reported but is given for the following years: 1879, 16,040 bales; 1880, 10,360 bales; 1881, 14,472 bales; 1882, 17,500 bales; and 1884, 23,140 bales. Regular lines of steamers sufficient for the business of the river ply between New Orleans and Shreveport, and statistics of their business are quoted.

At Shreveport the total receipts of cotton from all sources are given, and shipments by river, as follows:

ARTICLES.	1879.	1880.	1881.	1882.	1883.	1884.
Receipts of cotton...	103,660	93,580	77,448	not given	108,000	86,250
Shipments by river..	65,025	33,580	37,474	55,243	....	....

5. 1873, p. 664.

The business in cotton over the Red River, from Shreveport down, is here given:

	1878.	1879.	1880.	1881.	1882.	1883.	1884.
Cotton, bales	193,000	145,633	129,000	183,000	135,557	160,440	107,085
" seed sacks.... }	198,000	170,000	170,000	100,000	121,210	....	....

It is reported during two different years that more than 50,000 tons of freight of all kinds were shipped from New Orleans to the Red River. Besides cotton, as quoted, the steamers do a large business in cotton seed products, hides, cattle, sugar and molasses, besides general merchandise. It is not stated that the lumber business is large, either for manufactured lumber or in logs. Rafting is not quoted as a prominent item. There is nothing to show that business upon the river is increasing in consequence of the improvements, though there is no doubt that it is conducted with more security than formerly. At Shreveport in 1884, twenty-one steamers were registered, and a regular line of 13 steamers ran between Shreveport and New Orleans, besides independent ones. Nine steamers made one or more trips between Shreveport and Fulton.

An analysis of the amounts appropriated for the river and which cover the period from July 1st, 1872, to July 1st, 1886, show that the annual amount has been $43,357, which applied to 595 miles of river will average $73 per mile per annum. For the amount of business interested it would seem that no extravagance could be charged to this river.

OUACHITA RIVER, ARKANSAS AND LOUISIANA.

This river, which is the principal tributary of the Black, and by it of the Red River, presents perhaps the best example of a stream whose improvement is best insured simply by the removal of obstructions. At special points works of construction have been built, but it was found, in general, best to confine the work to the removal of actual and possible obstructions, snags, stumps, drift piles, wrecks and leaning trees. After using a crane boat for several seasons, an iron-hull snag boat, the O. G. Wagner, was built in 1878, and has been used since. The survey of the river shows that from Camden, Arkansas, to Trinity, Louisiana, 294.07 miles, the fall is 64.5 feet, with the average of 0.22 feet per mile. The depth of low water navigation has been increased from 12 inches to three feet, and high water navigation has been bettered. The records of the commerce of the river, which flows through a superb cotton region, are full and will be given principally as to that staple:

CALENDAR YEAR.	Cotton Seed	Cotton.	Steamers plying.
	sacks	bales	No
1871	....	151,458	...
1872	....	89,034	...
1873	....	103,679	19
1877	....	150,000	8
1878	656,620	131,324	19
1879	362,320	179,700	6
1880	....	128,707	19
1881	228,315	144,478	26
1882	....	116,544	...
1883	....	130,440	14
1884	94,700	97,678	16

Appropriations in 1871, 1872 and 1873 amounted to $211,000, and since then $79,000 more. The total of $290,000, covering from 1871 to 1886, will be $19,333 per annum, or $65 per mile per annum, to protect a commerce whose up freight is estimated at $2,000,000 per annum.

YAZOO, MISSISSIPPI.

This stream is formed by the Coldwater River, flowing into the Tallahatchie, and then by its junction with the Tallabusha. On the lower Coldwater, for 80 miles, work was done in 1880 and 1881; but not continued, as commerce did not justify further expenditure. Upon the Tallahatchie, for 190 miles, to the junction with the Tallabusha, work has been done since 1879; and upon the lower 90 miles of the Tallabusha, since 1881. To the the mouth of the river is 173 miles. During a portion of its course the Yazoo divides into two branches, one of which is Tchula Lake, which received attention first in 1881. The principal tributary of the Yazoo is the Big Sunflower, navigable in low water for 135 miles, and in high water for 280. The basin of the Yazoo covers about 13,800 square miles, and the navigable system of rivers give a total mileage of 813 miles at high water. When the war closed many wrecks were left in the Yazoo as a consequence, and the first appropriation of March 3d, 1873, of $40,000 was made for their removal, and paid for the clearance of the most important. Since that time operations have been confined to the employment of snag boats, and the removal of five wrecks. Commercial statistics are incomplete.

CALENDAR YEAR.	Steamers in use. No.	Cotton seed. sacks	Cotton. bales
1878	...	....	101,722
1880	...	125,000	95,135
1881	...	60,687	73,427
1882	...	55,000	69,000
1883	4	50,000	45,000
1884	6	75,000	50,000

There are reported to be shipped on the Tallabusha 12,000 to 15,000 bales of cotton per annum, of which 3,060 were sent by river to Vicksburg; in 1878-9, 3,500 bales. A weekly boat upon the Big Sunflower in 1875-6 took out 14,161 bales of cotton and 12,522 sacks of cotton seed. An analysis of appropriations made for these streams is as follows:

RIVERS.	Dates of appropriations	Total appropriations	Miles covered.	Per annum.	Per mile, per annum.
Yazoo, for removal of wrecks...	1873	$ 40,000	.....	....	.....
Yazoo..........	1875–1884	103,000	173	$9,363	$60
Yallabusha............................	1881–1884	9,000	80	1,800	22
Yallahatchie	1879–1884	24,000	190	3,429	18
Tchula Lake..........................	1881–1884	7,000	80	1,400	17
Big Sunflower........................	1879–1884	42,000	280	6,000	21
Coldwater (work abandoned)...	1879–1880	11,000	.....	....	.....
Total.................	....	$236,000			

While there is nothing to show that the total visible commerce upon this system of rivers has increased, during the progress of improvements; there is reason to believe that this sum, given for the navigable waters of a region greater than each of eight states of the Union, was well spent, when the total commerce of the thirteen years, from 1873 to 1886, is considered.

WHITE, BLACK, AND LITTLE RED RIVERS, AND THE SAINT FRANCIS.

These rivers which have been classed together at times in the various appropriation bills, drain the waters of southern Missouri and northern Arkansas. ⁶ White River rising in northwest Arkansas,

6. 1871, p. 366-358.

flowing east of north into Missouri, has been surveyed from the point, Forsyth, where it begins a southeasterly direction to its mouths, into the Arkansas and Mississippi. In this distance of 515 miles, only the portion from the junction with the Black, its principal tributary, 350 miles is fairly navigable. Above this point for 39 miles, there are difficulties; and between Jacksonport and Buffalo Shoals, 130 miles, are 63 shoals. Further on, the stream is little other than a mountain torrent. Black River, from Pocahontas to the junction, 151 miles, may be called navigable. Little Red River, upon which some work was once done, the next tributary, could not be used for more than 50 miles. The Saint Francis is a stream draining by its head waters southeastern Missouri, and flowing at low stages in its own channel. At high stages of the Mississippi it not only receives large volumes of overflow water from that river, but overflows itself into the headwaters of the Black, and thence into the White. Its traffic is limited, and in 1871 only one steamer was reported as doing a limited business in other than the cotton season, although 135 miles below Wittsburg, and 120 above it could be reached when water allowed. This entire region of country will grow cotton, and is supplied with few railroads. Steamboats have always been in use more or less, when they could be used, and are now. The area covered is large, probably over 20,000 square miles, and although only a small portion of this is accessible to the navigable portions, yet we should expect to find more commercial statistics reported than is the case. Work began on this system of rivers in 1871, the snag boats operating on the large rivers, doing the first work. From 1870 to 1884, twenty-four different items, from $5,000 to $40,000, and aggregating $380,000, have been given to these rivers. Although it is repeatedly stated that steamers on these rivers handle large amounts of cotton, in no instance are exact figures given.

A few instances have been selected, for examples, in which the reports show the expenditure of money to be a marked success, either in respect of improvement of navigation or of commercial benefits:

Between 1871 and 1874, the Roanoke River, North Carolina, received $45,000, expended in removal of obstructions and building of dikes; in 1882 to 1884, $8,000 more for repairs to this work. The Pamplico and Tar Rivers, North Carolina, received $35,000 since 1879, spent on 52 miles of river, to the benefit of commerce. Over 30,000 bales of cotton and much general merchandise shipped on the river in each of the years, 1883 and 1884. When work began navigation was confined to high stages.

The Neuse River, North Carolina, was practically closed to navigation when work began in 1878. The expenditure of $215,000 over 175 miles of river, developed a commerce of over 40,000 bales of cotton, besides shingles, rice and general merchandise, and 14,500,000 feet of lumber for 1884.

The Trent River, North Carolina, upon which work began in 1879, has received an extension to its navigable water of over twenty miles, and over 5,000 bales of cotton were shipped in 1884. Appropriated, $42,000; miles worked, 43.

Contentnea Creek, North Carolina, not navigable in 1881, $25,000 expended over 40 miles, prepared a navigable channel for steamboats drawing 3 feet, which carried 6,600 bales of cotton in 1884.

French Broad River, North Carolina, $43,000 since 1876, expended over 32 miles from Brevard, opened a channel 35 feet by 2½ feet; where before, the ruling depth was 1 foot, thus opening an inexpensive communication through a wild and difficult country.

Cape Fear River, North Carolina, below Fayetteville. Between this point and Wilmington, 113 miles, were expended in 1881, $10,-000, in the purchase of the right of taking tolls, exercised by a chartered company, and $55,000 between 1881 and 1884, in the removal of obstructions and the dredging of a channel 5 feet deep through Thames Shoal.

Commerce by river: In 1882, cotton, 7,954 bales, in 1883, 15,198; rosin, barrels, in 1882, 91,923, in 1883, 97,516; of tar, turpentine and spirits of turpentine, 40,573 barrels in 1882, 49,429 in 1883, and about 37,000,000 feet lumber in rafts each year.

The peculiar difficulty of opening navigation upon the Yadkin, which was, and remains obstructed with mill dams, limited work to 21½ miles, which now has a navigable channel through rock bars, but no commerce has been developed. Since 1879 has been expended $77,000.

On the Great Pee Dee River, South Carolina, 50 miles of river have been improved, by the amount of $27,000 since 1880, which has made navigation easier and safer. In 1883, 39,000 bales of cotton and 21,500 in 1884, are reported among the river shipments.

The Wateree River, South Carolina, has been worked over 45 miles from Camden down, securing a channel 60 feet wide and 4 feet deep, at an expense of $31,000 since 1881, but no commerce is reported. There are 96 miles of river to the junction with the Congaree, that can be improved and utilized.

The two Salkiehatchie Rivers, which become the Combahee, South Carolina, have had snags and obstructions removed, at the head of tidal influence, at an expense of $8,000, and to the interest of raft navigation.

Upon the Savannah River $57,500 was expended between 1881-4, in the removal of obstructions from the 248 miles between Savannah and Augusta. For the greater part of the year the river is navigable for boats drawing 4 to 5 feet, but during the dry season in autumn the water is very low at various places and boats could not reach Augusta.

Above Augusta the river has been improved in a distance of 64 miles, by the removal between 1880 and 1882, of rocky ledges, bowlders and gravel shoals, at an expense of $39,000; commerce does not appear to be responsive. No commercial statistics are reported for the lower Savannah.

The Flint River, Georgia, was only navigable over 36 miles at high water, when work began in 1878. Since then, at an expenditure of $97,000 on 70 miles, navigation is secured, which in 1884 occupied eight steamboats, which carried 21,760 bales of cotton and much general merchandise.

The Chattahoochee River, between Alabama and Georgia, was not navigable at low water when work began in 1874. Since then, the expenditure of $203,000, over 112 miles, has secured a steady business, which in 1884 employed eight steamboats, carrying over 47,000 bales of cotton and other merchandise. These two rivers form the Apalachicola River, Florida, which was obstructed with numerous snags before improvement began in 1884, and was completely blocked for a distance of 6 miles, beginning 50 miles above Apalachicola. Now the river is fairly navigable to a fleet of steamboats operating upon this system of rivers. Expended, $36,500.

The Oconee and Ocmulgee form the Altamaha River, the most important river lying entirely within the State of Georgia. Throughout its length of 155 miles, the chief obstructions are rock ledges running nearly across the river, sand bars and snags. In some places a permanent improvement has been effected, in others the operations of the snag boat should be continued longer. A channel 100 feet wide and 4 feet deep is secured for over 50 miles, at an expense of $35,000 since 1881. More than 50,000,000 feet of lumber are reported as going down this river for shipment from Darien.

Of the small rivers of Florida, the Withlacoochee received $10,500; the Caloosahatchie, $10,000; Peas Creek, $7,000; the Escambia and Conecuh, its tributary, $40,000, for removal of obstructions since 1881. Of these streams the last named have a commerce of some importance in timber, and steamboats with barges are used on the Escambia.

The Cahaba River, in Alabama, a tributary of the Alabama, has been worked over from Centreville to the mouth, 88 miles; expenditure, $30,000, since 1882; a small steamer was put upon the river and does a fair business. Railroad bridges across the stream are so built as almost to prevent any navigation, and should be altered.

The Alabama River was navigable with difficulty before work began in 1878. An expenditure since then of $130,000, over 374 miles, has materially bettered navigation. In 1883 47,052 bales of cotton, and in 1884 37,759 were carried on the river; which has also a rafting business of some importance.

The Black Warrior River, Alabama, was practically closed in 1875, when work began on it and the Tombigbee River. Since then $227,000 have been spent on the two rivers; over 140 miles of the first, and 416 of the second named. When work began on the Tombigbee an interrupted navigation existed over the upper 173 miles, in low water; and on the next lower 100 miles the navigation was not good.

The snagging has been completed over 278 miles, but will need annual revision. From Demopolis to Mobile, 243 miles, the river is ordinarily navigable throughout the year for boats drawing not over two and one-half feet. For 51 miles more, to Vienna, an increase of three months has been gained in the average boating season. Between Vienna and Columbus, 122 miles further, the average boating season has been extended two months.

There were shipped over the river 70,762 bales of cotton in 1883; and 41,088 in 1884; besides general merchandise. Above Columbus, since 1878, there have been spent $28,000 on 144 miles of river, securing a local navigation. Before improvement navigation was not possible over 79 miles of this stretch.

On the Noxubee River, Miss., navigation was impossible when work began in 1881, no steamboats having been on the river since 1859. Since 1881, an expenditure of $37,500, over 91½ miles of river, of which 40 miles was well worked, allowed steamboats to go to the county bridge at Macon, during high water, carrying 2,000 bales cotton, 3,500 sacks cotton seed, besides general merchandise, in 1884.

Of the net-work of small streams in Louisiana and Mississippi, and of interlacing bayous of the delta of the Mississippi, many are natural avenues of communication; some have deteriorated as such, from different causes, and some have been benefitted by government work.

On the Amite River, La., 40 miles were worked over at an expense of $13,000; Tangipahoa 41 miles, $9,000; Tchefuncte, $3,000; Tickfau 20 miles, $4,000; all having more or less commerce.

Bayou Courtableau, La., has a weekly line of steamers to New Orleans, and has received an expenditure of $19,000, in closing side bayous for the concentration of low water flow.

Bayou LaFourche, La., has a line of steamboats, and both banks are in a high state of cultivation, with sugar, and rice plantations. In the removal of snags, $30,000 have been expended. Similar work was done on Bayou Bœuf, La., $15,000 since 1881; and Bayou Barthol omew, La., and Tex., $18,000 since 1881; each stream being navigated by one steamer.

Tensas River, La., on which was spent $7,000 since 1881, had three steamboats plying on it in 1884.

PEARL RIVER, MISSISSIPPI.

[7] Perhaps no stream in the country has had its character so entirely changed in the last 50 years as this. Then, it was an excellent, navigable, clear water stream, and up to about 1860 while gradually changing its character, because of increasing cultivation of its bottom

---

7. 1879, I, p. 899.

lands, and of efforts to straighten its course, it yet remained one of the principal commercial arteries of the State of Mississippi. The stream is now as muddy as the Mississippi River; its length has been shortened fully one-tenth by cut-offs, and its banks in nearly every bend are annually caving. The planters have been driven by floods from their bottom lands, and the commercial value of the river destroyed by these attempts to shorten it. Planters are now forced to haul their products over clay hills to the railroads.

From Carthage to Jackson, 105 miles, the river is 120 to 250 feet wide, with a depth, at extreme low water, of fully 4 feet. The bed is of clay with sand and gravel and one rock bar. The stream is very crooked and winds through an alluvial bottom from 1 to 7 miles wide, heavily timbered with cypress. The difference between extreme low and high water is 18 feet at Carthage and 42 at Jackson. For 40 miles below Jackson are high bluffs, which confined the river so that a flood height of 57 feet was reached in 1874. For 210 miles further the bluffs are far apart. For a part of this distance the river bed is 300 feet wide, and the bed, rock, with banks 20 to 30 feet high above the rock. At 264 miles below Jackson is the head of a delta, where the river divides into two main branches; one on the west emptying into the Rigolets, and that on the east into Mississippi Sound, was, before the war, the main navigable channel. From the head of the delta down, the conditions resembled those of the Red River Raft region. Here the flood of 1874 was 26 feet above low water. Throughout the whole distance, thus briefly sketched, the cypress trees and stumps, logs, snags, and overhanging trees in immense quantities had driven away all navigation, in so much that estimates could hardly be formed of the cost which would be required to re-open the river. Between Carthage and Jackson sums amounting to $18,500 were granted between 1879-1882, and the river was made practicable for steamboats at a stage of the water which hitherto they could not have used.

Below Jackson, $80,000 were granted during the interval 1879-1884, and obstructions have been removed and the closure of small bayous begun for the concentration of the low water flow near the delta. Commerce is limited, but has begun to resume its former course down the river to New Orleans. The surveys covered 420 miles of river below Carthage.

BAYOU TECHE, LOUISIANA.

The commerce of this stream is probably greater than that of any of the same length in Louisiana. The lands bordering the bayou are very rich, and are all under cultivation, principally in sugar cane. It is in the heart of the sugar industry of the State. Cotton, cattle, hides, wool, moss, and lumber are produced in large quantities. The trade supports a line of steamers making regular trips to New Orleans, besides steamers making daily trips to Morgan City, and other small steamers in local trade. Under an appropriation of $6,000, in 1881, this bayou was cleaned of logs and snags from St. Martinsville to Leon's bridge, 36 miles; and with subsequent sums of $26,509 it was hoped to extend the low water navigation, by beginning a lock and dam five miles below Saint Martinsville. A second one was in contemplation, but neither would be built until an adequate sum was on hand.

If a canal be made to connect this bayou with Grand Lake, for which $25,000 was granted in 1881, nearly all of the commerce of Bayou Teche would use it. Bayou Teche is called by that name from its connection, or source in Bayou Courtableau, to the Gulf. In its lower course large steamers use it, but are not able to go on the upper course. Upon the lower portion an appropriation of $17,500, in 1870, removed the snags, and did marked good.

A careful examination of the records in the cases of the improvements of rivers of this class, in which the method of improvement adopted was the removal of obstructions, with only incidental use of contraction methods; and of which the majority of the instances have been summarized in this chapter; has given me the following synopsis:

There are fifty different rivers upon which work has been done; and they flow through one or more of the nine states from North Carolina to Texas. There are 6,684 miles of actually navigable water concerned; and $3,492,500 have been appropriated, or an average of $525 per mile.

This is to include the appropriations from 1870 to 1884. The Red River, of Louisiana, received the largest sum, $687,000; and only eight others received more than $100,000, during the entire period.

There is every reason to believe that this investment has been carefully made, judiciously expended, and with adequate results.

CHAPTER VIII.

THE MISSOURI RIVER.

[1]The Missouri River is the longest of any in the United States, and is, with the exception of the Ohio, the largest tributary of the Mississippi. From its source in the Rocky Mountains to its junction with the Mississippi, it is probably over 3,000 miles, and drains an area of 572,672 square miles.

It is navigable for nearly its whole length. Its tributaries are not of great size, though often of great length, and are rarely navigable.

---

1. 1881, II, p. 1650, Maj. Suter.

The country through which it flows is mostly one of small rainfall, and its really large discharge is due to the great area that it drains, and the mountain snows and ice of its head waters.

Its most salient and striking features are the remarkable impetuosity of its current, and slope, which is considerable for so large a stream. The rapidity of the current, and the general instability of the banks and bed, give rise to the excessive turbidity of its waters. It is, in fact, the greatest silt carrier in the country, and the enormous mass of sediment which it brings forward, forms the great bulk of that received by the Mississippi from its tributaries. Its influence upon the main river is most marked; indeed, it is its prototype in its main physical features, and from the navigation point of view, at least, it may be said to have a marked controlling effect upon the main trunk stream. Only that portion between Sioux City and the mouth, a distance of 781 miles, will be specifically considered. During the period of possible navigation, the range of water surface varies, in different localities and years, from 16 to 20 feet.

The amount of water discharged varies more than these limits would suggest. At St. Charles, in 1879, the discharge varied from 26,446 to 298,537 cubic feet per second, for a range of $17\frac{1}{3}$ feet on the gauge. It was further inferred that low and high water discharges would be about 15,000 and 430,000 cubic feet per second, or about as one to twenty-eight. At Sioux City this variation in discharge remained about the same, though the flood discharge was about 100,000 cubic feet less. Regular floods generally occur in April and June, the first extremely violent, and rarely lasting over a week or ten days. The June rise is generally higher, and of longer duration, being influenced by local rains and a general saturation of the soil. It is followed by a steady fall, until cold weather brings a sudden fall of three or four feet. There are occasional small rises of short duration. The rate of travel of the crest of floods is, on an average, about six miles per hour.

The main valley of the river consists of a great rock trough, cut down from the general level of the country to a depth considerably below the present level of the valley. The rocky banks form bluffs along the stream, and the bed is also of rock. The great trough seems to have been filled at one time with the glacial drift deposits, which also cover the adjacent country, and subsequently in part cleared out by the great river, that probably occupied it in early post-glacial times. The drift deposits seem everywhere pretty well sorted out; bowlders are generally found next to the rock, and deposits of varying degrees of coarseness above. The main feature is the great preponderance of extremely fine sand, which, with the addition of a very small amount of alumina, form an extremely tenacious clay, which is met with everywhere, and is formed in the bed of the stream wherever the current is unusually slackened. Large beds, or pockets, of very coarse gravel, or pebbles, are also met with in borings, at different depths below the surface. The valley below Saint Joseph averages about two and one-half miles in width; above that point it is wider.

The general depth of the rock bed below the surface of the valley varies from 70 to over 100 feet, wherever borings have been made, and no rock in place has been met at a higher level. At the few places where it is found in the bed of the present stream, spurs or ledges projecting from the main bluff are the cause, while the general level of the true rock bottom remains unchanged. If at any place rock did occur within the bed of the stream, it would be indicated by local increase of slope, but such is nowhere the case. At points on the river, layers of bowlders of considerable size form the bed of the stream, and have been mistaken for rock in place. At Saint Charles the piers of the bridge pass through a layer of this description, but the bed rock was found below the bowlders, and at the usual level.

The velocity of the current is very great. At low water this velocity is from 2 to 3 miles per hour, and in floods it amounts to 10 miles per hour or more.

Therefore, and because of the large amount of very light material in the bed and banks, the amount of bank erosion and scour, and fill of the bed is very great and very rapid. Bank erosion to the extent of 2,000 feet per annum, over long distances, has been noted, and to a greater or less extent it is constantly going on, even during low stages.

The enormous amount of material thus precipitated into the river, together with that scoured from the bed, causes the formation of innumerable bars, which, even at high water, obstruct navigation. These bars are constantly in motion, changing from day to day. The channels through them are also changing both in depth and location, rendering navigation correspondingly uncertain. Owing to the incessant bank erosion the river is constantly increasing the width between high banks, as bars are raised in height only after intervals of time, and are meanwhile liable to be swept away. Erosion occurring on the upper sides of points causes these points to move bodily down stream. When the erosion is more rapid than the compensating growth the point is swept away and the river dangerously straightened.

When a point or projecting neck is attacked on both sides, a cut-off is soon formed, which also acts detrimentally by increasing the local slope and inaugurating other destructive changes. The caving of the banks precipitates into the river countless trees, which form the snags that constitute, in the strong current, most serious dangers to navigation. The slopes vary from point to point, being, as usual, greatest on the crossing of the bars, and less in the pools above them. The general slope of the river, however, is remarkably uniform, and averages 0.′88 to the mile, both at high and low water. Owing to the great movement of the sand, when the current is strong, the bars are rapidly built up as the river rises, and although elsewhere the river follows the general law of scouring its bed at high water, and filling it at low, yet at these places of local engorgement the depth over the bars, as also in the channel leading through them, is much less than the low water depth plus the rise of the water surface. The actual navigable depth of the river varies from about 3 feet at low water to 9 feet at high water, though at the latter stage the channels are wider and the number of very shoal bars reduced.

The amount of sediment carried by the water is very great, this amount increasing with the volume of discharge. From observations carefully made at St. Charles, the proportion by volume, of sediment to water, was as 1 to 424; and by weight, 1 to 265. During the entire year the amount of sediment passing St. Charles would have covered a square mile to a depth of 197.58 feet. The increase in amount of sediment from surface to bottom was such as to lead to the inference that the amount moving in the lowest stratum might be fully equal to the amount actually collected. A just idea of the amount of solid matter brought into the Mississippi by the Missouri annually, can be based from these observations; and the total amount is estimated to be 11,000,000,000 cubic feet, or enough to cover a square mile to a depth of 400 feet. The water is so heavily charged with sediment that decrease in velocity is immediately followed by a deposit, but the converse of scour following an increase of velocity, although apparent, is not so well marked nor so extensive.

Wherever certain limits of width are exceeded, the river very generally shows the narrow pools, or channels of deep water, one adjacent to each bank, closed at each end, by the bar which, stretching down stream, lies between the pools. It is generally impossible to pass from one pool to the next on the same side of the river, but a crossing is effected through one of numerous channels by which the flow is made over the bar. The general dimensions of the bars depend upon the width of the river. The exterior boundaries follow the outlines of the main shore, and the bar joins the shore between the foot of one pool and the head of the next.

While the general position of the bars remain constant, their shape, the depth over them, and the position and depth of the channels through them are constantly varying.

These bars extending as they do entirely across the stream, act like low or partly submerged dams, and thus cause pools in which the slope is much below the average, while on the crossing from one pool to the other, it is much above the average. Even at high stages this difference of slope is very conspicuous.

At many points along the river are found reaches of moderate length where the width is too small to allow the formation of a middle bar. From the date of the earliest land surveys (1817), these places seem to have remained almost unchanged, there having been very little erosion. The slope is usually less than the mean slope of the river.

The preceding clear description of this river, although abridged from Major Suter's fuller account, will suffice for the present in following the course of works of improvement, and will be referred to hereafter in connection with the lower Mississippi.

IMPROVEMENT.

Beginning with 1868, regular snagging operations have been conducted upon the river. During the earlier years when commercial interests were of some importance, from one to four boats were employed annually for longer or shorter periods. The largest year's work reported was that of 1872, when four snag boats removed 2,226 snags, weighing 24,412 tons, besides 72 drift piles; cutting down 12,101 trees, and running in all 3,596 miles to do this work. The smallest year's work was that for 1877, when one steamer removed 161 snags only. The funds for the work came from the general appropriation for the three great western rivers, the Mississippi, Missouri and Arkansas. This work has been continued every year, but of late years is generally confined to the lower part of the river.

The act of 1875 ordered a survey made of the vicinity of Saint Joseph, Missouri, where the erosion of a point threatened to make a cut-off in such a manner as to leave the railroad bridge, crossing the Missouri at that point, upon dry land. This erosion and a similar shifting of the channel at Nebraska City, which threatened to destroy the water front of the city, besides damaging other property in the neighborhood, caused the first work attempted towards controlling the current and protecting the banks of the river. Allotments of $10,000 and $15,000 were made in 1877 out of the general appropriation of August 28th, 1876, for the Mississippi, Missouri and Arkansas Rivers, for these places. It was proposed to secure deposits of sand by placing light floating weed dikes in places and to protect the eroding banks by brush mattresses.

At Saint Joseph work began August 8, 1878, and 10,240 linear feet of bank were revetted. But a complication was in view here, because a cut off was threatened 20 miles above, which would shorten the river 4½ miles. This afterwards occurred, and work for ensuing years was continuous and troublesome in revetting necessary places and repairing damages which were the consequences of ice and floods. As the railroads were the interests at stake, they did much revetment work themselves, and as a result of the whole, the erosion was controlled by the expenditure of $109,000 up to 1882, when the system of operations was modified as will be shown.

At Nebraska City floating dikes were first put in, and with good effects; later on, much revetment was done at different points, and held with great difficulty and expense. Afterwards the ice and floods, particularly in 1881, carried away nearly all of the old work, causing an entire revision of project.

In 1882, buoy screen wire dikes were placed in position, and successively carried away, although a portion being retained promised better results. Upon the whole, however, the expenditure of $52,000 during the four years preceding the absorption into the general plan, cannot be considered successful.

The act of June 18th, 1878, contained items for these places, and for five others besides. At four of these, Fort Leavenworth, Atchison, Omaha, and Council Bluffs, the interests to be protected were the railroads; who were thus quick to take advantage of the opening presented; and at only one, Sioux City, was it even pretended that navigation interests were to be consulted.

The following year, the act of March 3d, 1879, included two more railroad bridges, at Glasgow, and Kansas City, besides two river points, Cedar City, Missouri, and Vermillion, Dakota, in addition to the items of the previous year. The remaining railroad bridge, Saint Charles, Missouri, appeared in the act of the next year, June 14th, 1880; and all were henceforth included. In 1881, therefore, work was done in accordance with the terms of the act, at seven points where the safety of railroad bridges were the interests involved; at one point, Plattsmouth, where railroads and farming land were at stake; at one, Brownville, Nebraska, where ferry and transfer interests were involved; and at five, Cedar City, and Lexington, Missouri, Nebraska City, and Sioux City, Iowa, and Vermillion, Dakota, where navigation interests, steamboat landings, or the regimen of the river, were avowedly under concern.

It must be recalled that during this interval it was the practice of the War Department, if not the law, to refrain from an expression of opinion as to the necessities for specific items, or, in fact, any discussion from any point, except as to how to secure the object aimed to be done.

During the years of active operations, from 1878 to 1882, the record of engineering methods and devices is full, ingenious, and instructive. The aim at all points was the same; to guide the current from an injurious to a salutary direction; to build up new banks, and bars where wanted; and to prevent the erosion of banks at bad places, where interests were threatened, or where the results would be bad upon the river itself. Dikes were built in an endless variety, from mere ropes with weeds tied to them, and anchored, to elaborate continuous wire-meshed screens, supported and anchored. Revetments of brush, put together in many ways, and placed in position after many fashions; until continuous wide wire thin mattresses were made so as to be handled and placed with great rapidity. Piles were used freely, and all the details of their best employment studied. Hydraulic grading of the banks, in connection with revetment work, was made easy and successful.

But the instability of the banks, the great power of the current at all times, and the added danger from ice and floods, make the whole work a gallant but almost useless struggle.

Finally Major Suter, believing that security was only possible by working in such a way that one system of works could protect another, urgently recommended in 1881, that it would be better to undertake with adequate funds the general improvement of the river, by which the whole valley would be benefitted and a good navigation assured. This course would, he thought, be in the long run much more economical, both of time and money, and would give tangible results at a very moderate cost.[2] Adopting this suggestion, Congress passed, in the act of August 2d, 1882, "For the improvement of the Missouri River from its mouth to Sioux City," the sum of $850,000. This is the close of the first period of this work.

The plan of work proposed by Major Suter was as follows: The ruling idea is to contract the river bed to such limits as will insure stability of regimen and uniformity of slope, width, and depth, at all stages, while at low water a maximum channel depth of twelve feet may be expected. Incidentally it is hoped, that when the improvement is completed a decided lowering of the flood line will take place. The general revetment of caving banks will prevent the destruction of immense amounts of most valuable farming lands. The methods to be followed will consist in reclaiming, by deposits from the river itself, such portions of its bed as lie outside the regulated channel way, with the expectation that these deposits will ultimately be built up to the normal height of the banks and form the new shore of the regulated river.

These deposits will be induced by the use of permeable dikes, which will check the current of the river sufficiently to cause it to drop its load of suspended sediment, while at the same time, they by their construction, will not invite the destruction which has invariably attended any attempt at massive work. The old and also the newly formed banks will be protected from erosion by brush mattresses to low water mark. Above that point the banks will be graded to a flat slope and protected by a covering of brush and stone.

In an improvement of this kind, it would be obviously desirable, in the interest of navigation alone, to extend the work from the mouth of the river upward, taking say the lowest 100 miles first, and so on. In the present plan a deviation is necessary owing to the wording of the law, and the obvious necessity of preventing the entire loss of work, at present incomplete. It was decided, therefore, with the approval of the Chief of Engineers, to initiate the work on that portion of the river lying between Charleston, Kansas, and Lexington, Missouri, a distance of 186 miles, which has within its limits five of the places where work had been in progress. Allotments were made to points other than these:

Saint Charles, Missouri...$22,000
Glasgow   "   30,000
Nebraska City, Nebraska.. 33,000
Omaha   "   20,000
Sioux City, Iowa... 15,000

---

2. 1882, II, p. 173.

The following table incorporates the principal features of the work already done at the end of the year 1882:

STATEMENT SHOWING WORK AT SPECIFIC POINTS ON MISSOURI RIVER AND RESULTS—1876–1882.

CITIES AND TOWNS.	Character of Interest to be protected by work.	Kind of Works Attempted, Other than Repairs.	Results of Works of Construction.	APPROPRIATIONS AND ALLOTMENTS.						Work covered by general project.
				Aug. 28th, 1876.	June 16th, 1878.	March 3d, 1879.	June 14th, 1880.	March 3d, 1881.	Aug 3d, 1882.	
Saint Charles, Mo.	Railroad Bridge	Revetment and Dikes	Works carried away, renewed, some results				$25,000	$15,000	$22,000	
Cedar City, Mo.	Citizens of that place	" "	Works finally successful			$10,000	15,000	15,000		
Glasgow, Mo.	Railroad Bridge	Short Dike	Work damaged and inadequate			15,000	20,000	20,000	30,000	
Lexington, Mo.	Steamboat Landing	Short Dike	Successful in part				15,000	10,000		
Kansas City, Mo.	Railroad Bridge	Dikes	First work much damaged, next better		$25,000	30,000	25,000	20,000		
Fort Leavenworth, Kas.	"	Short Dike	Much damaged			10,000	8,000	8,000		Work covered by general project.
Atchison, Kas.	"	Dikes			20,000	20,000	20,000	20,000		
Saint Joseph, Mo.	"	Dikes	Damage, but much work secured good results		50,000	9,000	20,000	20,000	20,000	
Brownville, Neb.	Ferry and Transfer Business	Dikes	All destroyed	$10,000	20,000		10,000	10,000		
Nebraska City, Neb.	Water front of City and Private Property			15,000	20,000					
Eastport, Ia., and Nebraska City, Neb.	vate Property	Revetment and Dikes	Nearly all old destroyed; new, better		20,000	30,000	14,000	20,000	33,000	
Platismouth, Neb	Farming Lands and Railroads	Dikes	Of no service				10,000	10,000		
Omaha, Neb.	Railroads and Private Property				30,000					
Council Bluffs, Ia., and Omaha, Neb.	erty	Revetment and Dikes	Much damaged, but secured in part, and revetment held		50,000	50,000	20,000	30,000	20,000	
Sioux City, Iowa	Navigation Interests and City Landing	" "	Stood well		12,500	50,000	8,000	7,000	15,000	
Vermillion, Dak.	Cut off Threatened	Revetment	Work failed			5,000	10,000	15,000		

[3]Under the new system of 1882, the seasons of 1882-3 were to be devoted principally to 186 miles of river, between Charleston, Kansas, and Lexington, Missouri, besides certain other points already under protection. The repair of old plant was an unavoidable necessity. This cost was great, and it was decided necessary to add new plant, to be prepared to carry on operations on a large scale.

Detailed items need not be given to show the number of barges, mattrass boats, hydraulic graders, quarter boats, pile sinkers and small boats prepared or acquired, but the reports assign $400,000 as the cost of the whole. The season of 1882 was employed at St. Charles in construction intended to preserve the status there. To close the slough behind St. Charles island a pile dike about 1,000 feet long was made. A sill of closely woven brush mattress, 80 feet wide, 12 inches thick, weighted with bailed rock, connected the high water banks. Through this a double line of braced piles was driven through the mid-stage section. This was finished by the middle of October. In St. Charles bend, grading for a revetment began October 2d, and when work closed, December 7th, 1,160 lineal feet of revetment had been placed, 135 feet wide, 12 inches thick. It had been intended to place 4,300 feet of this revetment, which would be terminated by a series of five spur dikes. Of these last, the first two, aggregating 1,000 linear feet, were built. As these dikes were located in a sharp gorge section, and under a concave bank, they required great strength. The piling could not be sunk more than 16 feet, so that a double line of A bents, braced laterally and longitudinally, and faced with a spiral net of galvanized wire, 36 inch square mesh, was constructed.

The break up of ice in the spring of 1883 was the most severe experienced, and proved disastrous to the work. The dike across the slough and the dikes in the bend were carried away entirely, and revetment so cut and torn as to require entire renewal. This revetment was again replaced to a length of 2,250 feet, and withstood the longest continued flood, and highest since 1844.

The allotment, however, made to this point, proved insufficient, and there was transferred from the Glasgow allotment $25,881, making a total for the year of $47,881.

In 1883-4 the revetment was completed by covering it with rock, thus exhausting the funds on hand.

[4]During high water in July, 1884, the river broke through the slough behind St. Charles Island, which has since become the main channel of the river. There will probably be no further cutting of the bank in St. Charles bend, though extensive changes are in progress lower down.

The total expenditure at this point, of which the object was solely to protect the approach to the railroad bridge, was $87,881.38; and the work has been useless, and the action of the river, apparently, uncontrollable.

---

3. 1883, p. 1302.   4. 1884. Part III., p. 1534.

At Nebraska City a revetment 5,100 linear feet of willow brush mattress 150 feet wide, and 12 inches thick, was put in at Eastport and protected with rock; no work done in 1884, and no loss reported; allotment of $33,000 not exceeded.

At Omaha extensive repairs made to the revetment, and 1,270 feet of extension made in 1883 and 1884. The work has held, and the allotment of $20,000 was sufficient.

At Sioux City the training dike was extended about 1,000 feet, but the heavy ice break-up in the spring of 1883, destroyed the dike entirely. The amount expended did not exceed the allotment of $15,000.

Upon the main stretch of the river selected for special improvement, the work done on the lower sixty miles was mainly in continuation of that already begun at Lexington, Missouri. Here a revetment 72 feet wide from the water's edge, 137 in total width, was put in for a length of 4,217 feet; and an upper bank protection of 3,119 feet. Construction ceased in January, 1883; and the heavy ice break-up in the spring caused no injury to this revetment, which received no additional work the following year.

Upon the next division of sixty miles the principal construction was in Kaw Bend, in continuation of the old effort to stop the erosion which threatened to cut off the railroad bridge by carrying the river to the north of it. Up to July 1st, 1883, about 7,350 feet low water mattress, and 7,610 feet upper bank mattress had been laid; besides extensive repairs made to the remaining revetment of 1879, about 1,200 feet in length. The ice gorge and break-up of 1883; the heavy wind and waves, and the high water of June, 1883, did not injure this revetment to any marked degree; but after that time a pocket began to cut in above and behind the head of the revetment, and during the course of the year flanked about 1,000 feet of the upper part, and was still in progress when last reported. Work was projected at Fort Leavenworth, but nothing done.

In the third division efforts were directed, at Saint Joseph, to repairing and extending the Elwood revetment, opposite the town, which had been much damaged, in 1881, by the destruction of dikes put in by the bridge company. About 4,100 linear feet of revetment were placed; similar in character to that referred to previously. At Atchison, during this year, an effort was made to rectify the channel above the bridge so as to bring it under the draw span at right angles. It was intended to construct a dike about 4,000 feet long, and certain cross dikes as were necessary. The construction of this was attended with difficulty, and finally suspended because of high water. Although not finished, this dike seemed to promise satisfactory results. The ensuing year, no work was done at either of these points, but no loss is reported, and the results are considered good.

The extensive surveys, rendered necessary, if a thorough knowledge of the changes in the river bed and its conditions, be required, cause large expense, which should be considered in estimating the cost of construction on the river, or at different points.

It does not appear from the record that any different character of work was adopted after the passage of the general appropriation in 1882, from that followed before.

It does not appear that any different localities were selected for the application of the system proposed, than those already selected; nor do definite results seem any more attainable, when funds are abundant, than previously. The officer in charge believes that the success of the improvement is problematical, unless the work can be carried on continuously with ample funds.

At this time, July, 1884, the system was again changed, and by the act of July 5th, 1884, a Missouri River Commission was constituted, similar in character to the Mississippi River Commission, an appropriation of $600,000 made, and Major Suter appointed the President of the Commission.

The Missouri River Commission, constituted as stated, made its first report under date of December 9th, 1884. After a general review of preceding events connected with the river improvement, a summary of the items of appropriations from 1876 to 1881, included, closes thus :

"The aggregate of these appropriations is large, amounting to $861,000, but no useful results, at all justifying an expenditure, were obtained or could be obtained under the system. The means were inadequate at nearly every point, nothing could be finished, and the incomplete work was an easy prey to the destructive forces of the river." The report then refers to the appropriation of 1882, as one by which the systematic improvement could be rendered possible. "Under this, some of the more important of the works previously undertaken were continued, but a very large proportion of it was devoted to the preparation of the machinery and other plant required for conducting the works upon a large scale." With reference to work under the new appropriation of 1884, the Commission approves and adopts the plan of Maj. Suter, already given, and the methods proposed for carrying it out, subject to such modifications and changes as future experience may dictate, but while expressing an opinion that the work is physically practicable, the Commission states that no estimate of its cost can be given. "The experiments heretofore made have many encouraging features. Their want of cohesion is sufficient explanation of their want of success. The trial, to be demonstrative, must be undertaken with ample means, and followed up without intermission for several years, over a continuous piece of river, the length of which must be considerable; while during the experimental stage, annual appropriations of $1,000,000 are recommended." The Commission admits that the cost of protecting any portion of the bank from erosion is so great that it is only in exceptional cases that it is justified; and that the amount of commerce upon the river at the time is small. Finally the Commission selected Kansas City as an initial point; 386 miles above the mouth of the river; and proposed to make the improvement continuous, working down stream, applying all the means placed at its disposal, as far as possible, to this purpose, protecting land and building up new banks as this becomes necessary for the preservation of the channel. A cut off threatened to be caused near St. Joseph, was an exceptional locality, justifying an exception to the proposed plan. No work has been done under this plan prior to 1885.

Of the commerce of the Missouri little is given in the reports. During the season of 1867, 28 cargoes were cleared from Saint Louis for points on the Upper Missouri amounting to 8,094 tons. From St. Louis to Fort Randall, from February 1, 1867, to October 1, 1867, were 212 clearances and 169 arrivals. In 1868 it was reported (p. 662) that the total arrivals and departures at Saint Louis, to and from the Missouri River for 1867, was 246; and for 1868, 286. In 1875, Maj. Suter states that upon the completion of the railroad to Sioux City the greater portion of the mountain freight was shipped from that point, and that the fleet of steamers rarely came below. More recently the completion of the Northern Pacific to Bismarck gave another starting place. From that point supplies were largely shipped, but when the railroad was further extended, less use was made of the river. Indeed, since 1875 the reports hardly allude to navigation, and never give the statistics found in connection with the improvements of other rivers. It was stated upon the floor of the House of Representatives at the last session (Record p. 2149) that members from Missouri and Iowa said that the river is of no commercial use, nor has it been for years, especially from Council Bluffs to the mouth. On the contrary, it was stated that one or two lines of packets plied daily on the river, making trips between St. Louis and Kansas City two or three times a week. A gentleman from Kansas was quoted on the floor as stating that the draw at Atchison had been opened but eight times for boats during the year.

Above Sioux City no continuous survey had been made, except over the 371 miles to Fort Pierre, and no work of construction except that at Vermillion, Dakota, begun in 1879, and discontinued in 1882.

Between Fort Pierre and Carroll is a long reach of river estimated at about 1,145 miles, of which examinations have shown that in its downward course the river gradually assumes the sandy and shifting character, and earns the title of the "Big Muddy," which so well distinguishes it at Sioux City. No work has been done over this reach. Above Carroll the river flows over a bed of rock and gravel. In this rocky portion work can be carried on at detached points with the expectation that each obstacle removed will be of permanent benefit to navigation; and all work of this nature has been done between Fort Benton and Carroll, 160 miles. The removal of rocks from the channel, and the building of wing dams at the most prominent points,

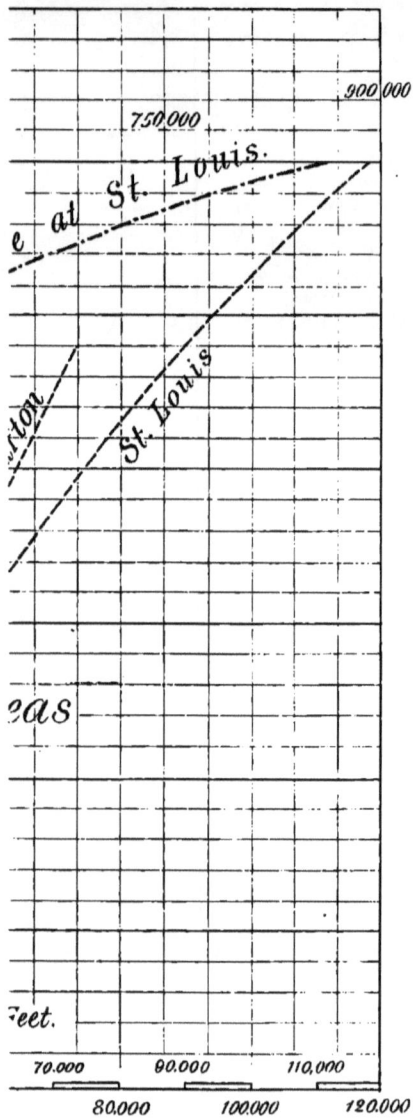

have been in progress since 1876, for which the total appropriation has been $450,000. Previous to the commencement of work in the river, boats could not ascend further than Cow Island during low water season; now they can go 84 miles higher, and the season is extended two months. The steamboat business is on the increase, a healthy down stream and local trade developing under the influence of the rapidly increasing population of the country tributary to the river.

CHAPTER IX.

THE UPPER MISSISSIPPI RIVER.

By the junction of the Minnesota River with the upper Mississippi, a river of great magnitude is formed, draining an area of 32,000 square miles. The junction of the Saint Croix River increases this area to about 39,600 square miles. At this point the lowest recorded discharge in winter is 3,418 cubic feet per second; and in summer the smallest discharge 8,533 cubic feet, the gauge reading 2.5 feet higher. By reference to the Ohio it will be seen how much these figures exceed those. The exact surveys of the Mississippi River Commission, and prior surveys, under the Engineer Department, give in the published reports, more exact knowledge of this than of any other of the rivers already studied.

The following table gives the principal figures of the survey of the river; the discharge observations being shown on the plate.

UPPER MISSISSIPPI RIVER DISTANCES, SLOPES AND DRAINAGE ARES.[1]

PLACES.	Distance from St. Paul. MILES.	Low water elevation above sea. FEET.	Slope per mile bet. po'ts. FEET.	R'nge bet. high and low water. FEET.	TRIBUTARIES. AREAS DRAINED Of Str'm. MILES.	TOTAL MILES.	Right or left bank.
Saint Paul Bridge............	0	678.00	......	19.72	......	19,903	....
Minnesota River..............	......	......	......	......	12,119	32,022	R
Newport.....................	8.50	673.51	.528	18.38	......	......	....
Hastings....................	27.	663.75	.527	18.76	......	......	....
Saint Croix River...........	29.	......	......	......	6,568	39,590	L
Dimond Bluff................	42.75	659.67	.258	......	......	......	....
Cannon River................	45.	......	......	......	1,039	41,229	R
Vermillion, Rush and two smaller streams.........	32-55	......	......	......	588	41,817	....
Red Wing	52.25	657.75	.202	16.60	......	......	....
Waconta.....................	59.25	656.40	.138	13.89	......	......	....
Chippewa River..............	82.	......	......	......	9,602	51,419	L
Beef River............	94.	......	......	......	452	51,871	L
Reed's Landing..............	83.85	656.30	......	......	......	......	....
Wabasha.....................	86.50	654.85	.557	12.05	......	......	....
Alma........................	95.50	649.48	.597	12.62	......	......	....
Zumbro River................	91.	......	......	......	1,366	53,237	R
Whitewater, Eagle and Rolling Stone Creeks.......	92-107	......	......	......	666	53,903	....
Minneiska...................	105.50	643.19	.629	......	......	......	....
Fountain City...............	117.25	637.41	.491	......	......	......	....
Winona......................	124.75	632.44	.663	......	......	......	....
Trempeleau River............	134.	......	......	......	700	53,608	L
Trempeleau	137.50	627.10	.419	15.97	......	......	....
Dresbach....................	148.	623.39	.346	......	......	......	....
LaCrosse....................	156.	621.23	.280	15.47	......	......	....
Black River.................	146.	......	......	......	2,880	56,483	L
LaCrosse River..............	155.	......	......	......	463	56,946	L
Root River..................	160.	......	......	......	1,685	58,631	R
Three small streams.........	163-177	......	......	......	389	59,020	....
Brownsville.................	166.	615.65	.560	15.76	......	......	....
Genoa.......................	178.75	610.28	.447	......	......	......	....
Victory.....................	187.	608.64	.490	......	......	......	....
Upper Iowa River............	187.	......	......	......	1,375	60,395	R
Lansing.....................	198.75	605.49	.260	......	......	......	....
Two small creeks............	224-228	......	......	......	652	61,047	R
Lynxville...................	211.25	601.76	.286	20.28	......	......	....
Allamakee...................	220.85	598.50	.333	......	......	......	....
Prairie du Chien............	227.02	597.57	.168	......	......	......	....
Wisconsin River.............	230.50	......	......	......	11,850	72,897	L
Clayton.....................	239.	594.67	.239	20.31	... ..	......	....
Glen Haven.	245.09	593.00	.273	......	......	......	....
Turkey River................	255.	......	......	......	1,679	74,576	R
Cassville...................	256.34	588.24	.390	20.34	......	......	....
Nine small streams..........	269-296	......	......	......	1,157	75,733	....
Waupeton....................	263.44	585.60	.425	......	......	......	....
Well's Landing.....	271.32	583.10	.316	20.07	......	......	....
Dubuque.....................	284.51	578.22	.370	21.73	......	......	....
Deadman's Bar...............	295.12	574.92	.316	......	......	......	....
Galena River................	301.50	......	......	......	1,895	77,628	L
Goldens	313.16	569.10	.344	......	......	......	....
Makoqueta River.............	317.	......	......	......	2,209	79,837	R
Apple, Rush and Plum Rivers..	321-329	......	......	......	610	80,447	L
Savana	327.	564.37	.258	......	......	......	....

UPPER MISSISSIPPI RIVER DISTANCES, SLOPES, ETC., CONTINUED.

PLACES.	Distance from St. Paul. MILES.	Low water elevation above sea. FEET.	Slope per mile bet. po'ts. FEET.	R'nge bet. high and low water. FEET.	TRIBUTARIES. AREAS DRAINED Of Str'm. MILES.	TOTAL MILES.	Right or left bank.
Fulton	346.11	559.74	.356				....
Comanche	354.61	556.50	.376	23.87			....
Wapsipinicon River	260.				2,957	83,404	R
Cordova	363.61	554.60	.205				....
LeClaire	368.86	554.	.127	13.00			....
Rock Island	383.61	533.77	1.372	18.40			....
Rock River	386.50				9,692	98,096	L
Buffalo	394.61	530.36	.309	17.17			
Fairport	407.24	526.72	.290	17.42			
Muscatine	415 62	523.60	.250	14.98			
Port Louisa	419.12	519.81	.356				
Iowa River	436.				12,385	105,481	R
New Boston	437.37	517.63	.264				
Keithsburg	442.87	516.80	.136				
Oquawka	456.12	509.82	.533				
Edwards River	438.				638	106,119	L
Henderson River	463 62				4,753	110,872	L
Four small streams	442-476				469	111,341	....
Kaintucks	461.24	507.60	.425				
Burlington	467.68	505.10	.388	15.00			
Skunk River	474.				4,753	116,094	R
Dallas	480.02	501.	.440				
Fort Madison	488.86	496.70	.205				
Montrose	497.71	494 50	.256				
Keokuk	508.83	472.33	2.009	20.29			
Des Moines River	511.33				13,692	129,786	R
Four small streams	493-525				1,527	131,313	....
Warsaw	512.87	469.70	.600	22.12			
Dodds	523.47	464 66	.493				
Tully Island	527.75	462.71	.456				
LaGrange	535.89	466.53	.759				
Quincy bridge	544.60	453.80	.399	21.06			
Fabius River and two creeks	550-555				3,086	134,399	R
Hannibal	563.41	444.95	.452	21.35			
Saverton	570.28	441.30	.376				
Mundy's Landing	579.66	437.78	.580				
Salt River	589.				2,815	137,214	R
Five small streams	591-626				282	137,496	R
Louisiana bridge	591.53	430.97	.493	20.50			
Clarksville	600.96	426.93	.428	23.01			
Slim Island Foot	611.80	420.90	.510				
Fulmouth	619.58	417.64	.483	23.75			
Sterling	623.59	415.74	.462				
Turner's Landing	628.14	413.49	.505				
Cap au Gris	633.50	411.04	.462	22.26			
Cuivre River and two creeks	638.				1,607	139,103	R
Barrack Island	640.04	407.83	.491				
Phelps Landing	650.60	402.54	.501				
Grafton	656.73	399.44	.505				
Illinois River					27,963	167,066	L

1. Bridging Mississippi River, Warren p. 31. Report 1880, II, p. 1522.

[2]From Saint Paul to LaCrosse, 156 miles, the course is winding, and the bed generally sandy. After every rise the sandy bottom changes its shape, and at low water presents serious, and sometimes impassable obstacles to navigation. An average low water depth of three feet is found on the bars. Large steamers are generally able to reach Prescott, during low water. When the river begins rising the sand bars are apt to shift; old channels filling up, and the bottom flattening out. When the water falls, the stream is spread over so great a width of river, that it is very shallow. In the course of time the water cuts for itself a new channel, and navigation is, again, soon restored. Changes similar to these often occur several times in the course of a season. An island in midstream is nearly always, at its lower end, a source of trouble. Here, two currents meet, the greater part of the water and the channel, cross the river from bank to bank, at a sharp angle, with a shoal as a result; the water spreading in thin sheets over the bars, while deep water is off the outer edge of the reefs.

The water of the upper Mississippi is clear, and no mud is deposited from it. Observations made during a period of a year show that 56 parts in 100,000 at Prescott; 7 at Winona; 15 at Clayton; 129 at Hannibal; and 108 at Grafton, are maxima; while at Saint Louis, for the same period, 336 parts were given, as the corresponding maximum amount of sediment found in the current. [3]The bed and bottom are nearly all sand. The bottom lands are overflowed many feet and their height is just that, to which the current will raise the sand as it passes over the bed; and is the same height as the tops of the highest bars, which differ from islands and bottoms only in not having been formed long enough to have produced a growth of trees. This sand is very heavy, weighing from 105 to 118 pounds per cubic foot, composed mainly of silica, with some iron. When stirred up it sinks again very rapidly. It is moved along mainly at high water, and stops as soon as the current slackens; so at low water comparatively little sand is in motion.

From Saint Paul down, the river is filled with numerous islands, no less than 526, having been numbered in the main river, above the mouth of the Illinois, besides great numbers not so enumerated. Between the bluffs, the whole bottom is interlaced with a net-work of sloughs, and water courses, carrying more or less water, at high stages of the river. Many of the larger islands are fine grazing, or farm land; but in general, they are abandoned to the luxuriant growth of willow, or the original forest. At low stages the velocity of the current in the river, is gentle; between two and three feet per second, or at the rate of three to four and one-third miles per hour. Velocities increase with the gauge reading, generally in a simple ratio; but which varies with localities. At Hannibal the increase in velocity is more rapid, as the river rises, than is the case at the upper points; or than at Grafton.

---

2. 1875, II, p. 456.   3. 1868, p. 318.

The curves of discharge are very similar at all points, but from the head of the river to Saint Louis, the ratio of discharge to gauge, gradually increases; slighter differences in high stages, making greater proportional discharges as the river widens, and increases in volume.

[4] In the bill of March 3d, 1867, an item of $96,000 was placed for two dredge and snag boats, and for their operation for one year upon the Upper Mississippi. The two boats were bought and operated, in scraping the bars rather than in dredging, and with manifestly good results where used; a depth of three feet being maintained. When not used for this purpose, the boats removed and cut up all snags and all overhanging and projecting trees, which were within the sphere of their operations. The relief afforded to navigators, during extreme low water, by these boats, during the season of 1868, their first experience, was marked. During the preceding fall those boats had had some preliminary experience, which gave opportunity to determine what adaptations would be desirable for the ensuing year. [5] In 1868 when work began for the season, the water was so low that large boats were stopped. Soon after the Caffrey began operations, however, the channel was deepened so as to allow their trips to be resumed. The river kept oscillating between a three and four feet depth upon the bars, causing work to be done over again after every rise and fall. During the season of 1869 low water again called for work early in July, and this prevented any suspension of navigation by the largest boats.

During the year it was recommended to close by dams the small channels, or sloughs, which drew off water from the main channel, as a necessary accompaniment to the temporary work of improvement in progress.

But little variation is noted in the execution of this work for several years. Each year the boats did what they could until in 1873 the Caffrey became too old to be of service, and was laid up, the other boat, the Montana, continuing at work. As a general thing very much assistance was given to navigation in pulling steamboats and barges off, when hard aground; and a channel depth of 3 to $3\frac{1}{2}$ feet was secured during low water by dredging; over the stretch of river on which time allowed operations to be conducted. By clearing the shores of overhanging trees, and those near the bank liable to cave into the river, the number of snags had been largely reduced so that in 1876 the whole number found between St. Paul and the Des Moines Rapids was only thirty-seven.

For the ten years from the date of the first appropriation a total of $352,640 had been appropriated and allotted, with which two steamers had been bought and operated.

---

4. 1868, p. 317.   5. 1869, p. 189.

The following are the general features of the work :

	1868.	1869.	1870.	1871.	1872.	1873.	1874.	1875.	1876.
Steamers in operation.......	2	2	2	2	2	1	1	1	1
Miles run...	5,904	.....		7,292	6,730	2,263	3,862	3,535	1,184
Snags extracted............	329	475		465	550	3	13	38	37
Leaning trees removed......	344	595		656	2,550	16	45	169	3,136
Stumps extracted...........	...	......		33	722	2	2	9	.....
Steamboats, barges & rafts pulled off bars........	.....	.....		33	10	11	1	18	2

In 1871, 1,600 feet of wing dams were built, and in 1873 were built dikes at four of the worst bars; dams in 1874, 5, and 6, at Pig's Eye and Nininger Slough; in order to concentrate the water in low stages to one of the two channels formed by the islands at these points. This point may be considered the end of the first stage of improvement operations.

The Montana closed her career at the end of the season of 1878, after fourteen years' service, and such parts of her as could be utilized were used in the construction of her successor, the General Barnard. This steamer was completed April 11th, 1879, and began work April 23d. The steamer is side-wheel, 218½ feet long, 37 feet beam, width over all 65 feet, with side wheels, 25 feet diameter, 12 feet buckets; boilers of steel, 3 in number, 42 inches diameter, 22 feet long, with ten 6-inch return flues in each; cylinders 20 inches diameter, 6 feet stroke; draft light, 31 inches. Total cost, $20,963.22. Is of very strong construction, furnished with two steam capstans, derricks, and other appliances, fitting her for the service in which she is engaged. This boat has since then continued in service and rendered valuable work in facilitating navigation. It is considered that without this assistance navigation on the Upper Mississippi could not exist. Additional amounts granted in later appropriation bills raised the sum already quoted to $517,140, as the grand total for this purpose to cover all expenditures from March 2d, 1867, to July 1st, 1886 ; out of this, three steamers have been bought and one operated constantly for every season beginning with 1868, and one other for five seasons. The average of $27,217 per annum for this service, upon this great river, would seem to be a wise investment, particularly when the commerce at stake is considered.

The necessity of closing as many side channels and sloughs as diverted water from the main channel was early recognized—Major Warren, in his first reports, suggesting the combined use of brush and stone for this purpose. And, as seen, when circumstances and means permitted, a beginning was made at once. But to carry out such a work called for more funds than had as yet been granted. By 1876 a plan of operations was digested, and fully reported, including a list of bars between LaCrosse and St. Paul, with estimates for the necessary dams. [6] Forty-three were enumerated in the 156 miles, and an estimate of $200,000 furnished for their construction. The bars upon that part of the river were the most numerous and troublesome. No estimate was made for several bars, the water at the time of the survey being good. Some of the worst bars, for a season or two, might not produce trouble, although difficulties were apt to arise in time. At this time special appropriations had been made for the harbors of Fort Madison and Burlington, where rip-rap dams were begun, and for Dubuque, where a sand bar was to be dredged away, and afterwards a training wall proposed. The general scheme of improvement was incidentally approved, by a board of engineer officers, as an essential to permanent improvement, to close all secondary channels in the Upper Mississippi [7] and concentrate the whole flow in a single channel.

As a part of third subdivision of the Mississippi transportation route to the seaboard, the whole subject is discussed, [8] and an estimate submitted for the entire distance from St. Paul to the DesMoines Rapids as $617,393.01. From DesMoines Rapids to the mouth of the Illinois River, at this time no survey had been made, nor any operations carried on, other than the removal of snags.

The act of June 17, 1878, gave $250,000 for the improvement of the Mississippi between St. Paul and DesMoines Rapids, and $100,000 from the latter to the mouth of the Ohio. The first works undertaken were at the localities where the greatest difficulties to navigation had been experienced. The dams were built of mattresses of brush, weighted down with stone. The lower layer of brush extending ten feet down stream beyond the layer above, succeeding layers were stepped back up stream three feet each. Except on the apron and top covering, where stone as heavy as a man could lift was used, the ballast was composed in part of stone as small as coarse gravel. [9] The following table illustrates the principal work done during the season in the upper river. The closing dams were assisted in this work by ten of the spur dams built. The nineteen other spur dams were intended to contract the water flow at wide, or bad places, to secure depth :

6. 1877, I., p. 529.   7. 1878, I., p. 720.

8. 1876, II., p. 176.   9. 1879, II., p. 1119.

	Miles from Saint Paul.	DAMS.			Shore protection, linear feet.	MAT'RIAL US'D		COST.	
		Spur. No.	Closing No.	Linear feet		Rock, cubic yds.	Brush. cubic yds.	Average per cubic yard	Total.
Pig's Eye......	3	10	..	3,575	1,200	8,883	5,630	$1.32	$18,573.32
Newport.......	8½	1	1	775	350	2,934	1,277	1.42	5,952.61
Hastings	27	4	..	1,700	400	4,681	3,569	1.15	9,475.63
Prescott........	30	2	1	1,695	800	5,127	3,991	1.13	10,359.77
Crat's Island..	86½	2	1	2,450	400	14,689	8,091	1.15	26,298.32
Beef Slough..	89½	3	2	3,200	700	9,859	6,485	1.23	20,069.81
Rollingstone..	115	2	2	2,280	800	7,755	4,423	1.23	15,087.85
Betsy Slough.	118	2	1	1,755	3,152	12,592	5,209	1.19	21,184.51
Queen's Bluff.	141½	2	..	2,010	....	4,926	2,549	1.50	11,212.24
Bellevue.......	314¼	...	2	1,350	350	5,562	6,549	1.33	16,161.97
Dallas..	494½	1	1	1,300	950	3,134	4,033	1.17	· 8,350.08

Below the Des Moines Rapids it was found necessary to make extensive surveys, over the whole stretch of river, in fact, before work of construction could begin. As soon as the maps could be made, plans for work at Gilbert's Island, and Slim Island, were submitted to and approved by the Board of Engineers for improving the Mississippi River. Beginning construction so late, the projects were carried out only in part:

	Miles from Saint Paul.	DAMS.			Shore protection, linear feet.	MATERIAL USED.		COST.	
		Spur. No.	Closing. No.	Linear feet		Rock, cubic yds.	Brush, cubic yds.	Average per cubic yard	Total.
Gilbert's Island..	573	4	1	3,212	1,862	6,115	4,860	$1.07	$11,724.20
Slim Island......	610	1	3	1,760	1,295	7,744	5,049	1.04	13,319.60

The result of the study of this part of the river showed that much work would have to be done to improve the river where obstructions were found; and to rectify the channel and protect the river banks, where bad bars might be readily be formed. For the greater part of the distance both banks of the river are composed of soft material. As a preliminary estimate, the cost was placed as follows:

1st. For improving river where obstructions exist, $921,294.

2d. For maintaining and rectifying channel, $1,125,102.

It being thus rendered evident that a natural division of the Upper Mississippi was found at the Des Moines Rapids on account of the increased cost of operations below that point, it will be well to continue our examination of the upper portion to a conclusion before proceeding with the lower.

During 1880 were built twelve closing and ten spur dams, in continuation of work already begun at six different localities, and at one place, as an initiation, together with a large amount of shore protection. At six other points dredging, removal of rock obstructions, repairs and minor work is reported, the whole actual construction account covering $109,660.34.

The season of high water extended through the summer, and a good stage of water for the early part of 1881 gave the year a good record for navigation. The water continued high throughout the whole season. Except during a very brief period in the summer no difficulty was experienced. Throughout the year the largest class of boats were able to reach St. Paul, except for about two weeks, when a bar at Pine Bend, ten miles below, interfered with their progress. Active operations consisted in building twelve closing dams from Saint Paul to Hastings, thus nearly completing the closing of the secondary channels on that portion of the river. At six other points extensive operations in continuation of prior work, repairs, closing dams and shore protection, were carried on. At Burlington the troublesome part of the river there, already under improvement at Rush Chute, received a large expenditure in the closing of Shokokon Slough. Altogether during that fiscal year $207,340.75 were expended, of which the purchase of steamers, barges, &c., involved $65,781.83.

The close of the season of 1883 found the channel from Saint Paul to Hastings, with all the side sloughs, closed by dams with crests at a minimum height of 4½ feet above low water; and with a great many wing and spur dams to contract the channel at wide places.

Construction continued throughout the season under the general plan, the account showing an expenditure of $183,422.03, of which all was used in actual work, excepting $19,590, in purchase of barges.

In August, September and October, 1883, the river was quite low, reaching for a few days almost, if not quite, low water mark, yet steamers of the Diamond Jo line reached St. Paul throughout the entire season.

There having been no appropriation made in 1883, the construction account for the fiscal year 1883-1884 amounted to $80,768.95 only, providing for the usual class of work at various points.

The work having now progressed during six years, it was claimed to be no longer experimental, and Maj. Mackenzie, in charge, stated that the system, if carried out, would give a good and permanent channel from Saint Paul to the Des Moines Rapids for the largest class of Upper Mississippi steamers during the season of navigation.

The following are the expenditures from the beginning of improvement to July 1st, 1883:

	DISTANCE. MILES.	AMOUNT.
Saint Paul to Hastings	27	$181,035.36
Hastings to head Lake Pepin	32	46,291.52
Head Lake Pepin to Alma	36	170,044.02
Alma to Winona Bridge	29	158,353.51
Winona Bridge to LaCrosse Bridge	31	41,018.05
LaCrosse Bridge to McGregor Bridge	72	23,938.27
McGregor Bridge to Dubuque Bridge	59	24,752.41
Dubuque Bridge to Clinton Bridge	67	17,057.48
Clinton Bridge to Rock Island Bridge	40	
Rock Island to Keithsburg	58	6,645.31
Keithsburg to Des Moines Rapids	60	114,273.47
Surveys and meter work		52,665.40
Snagboats and wrecking		6,436.70
Plant, estimated value		54,294.06
		$906,805.86

To secure local objects, appropriations have been made, beginning with 1876, for points upon the Upper Mississippi, amounting in total as follows, and expended in accordance with the adopted plan of operations when possible :

Galena River and Harbor, dredging	$64,000
LaCrosse Harbor, closing dams	12,500
Guttenberg Harbor, closing dams	13,000
Dubuque Harbor, dredging	61,000
Rock Island Harbor, dredging	12,000
Andalusia Harbor, closing dam	6,000
Muscatine Harbor, dredging	20,000
Burlington Harbor, dredging and closing dams	30,000
Fort Madison Harbor, closing dams	24,100
Harbors of Refuge, Lake Pepin, piers	35,000

The work at several of these points has been of manifest value to the general navigation of the river, and there is reason to believe that in all cases local commerce has been benefitted by the work done.

Below Des Moines Rapids no difference exists in the plan of improvement adopted, or in the method selected ; but the increased dimensions of the stream require all operations to be upon a large scale, and with increased expense.

Upon the upper stretch, from Keokuk to the Quincy bridge, the expenditures from July 1st, 1878, to July 1st, 1884, were $80,653.58, the distance being 40 miles.

The closing of Canton and Smoot Chutes, together with subsequent repairs, involved 12,581 cubic yards rock, 5,492 cubic yards brush, besides gravel, at a total expense of $25,398.10. Operations at Gregory's Landing consumed 17,445 cubic yards rock, and 14,973 cubic yards brush, at a cost of $40,515.20.

Extensive revetment and spur dams, at Quincy, and Hannibal, are the principal operations in the second stretch of 25 miles, from Quincy bridge to Saverton, and the total expenditures in the distance $49,352.76.

At Gilbert's Island, 11½ miles below Hannibal bridge, was the practical head of navigation in 1879; and the control and concentration of the river proved to be a most expensive and difficult operation. The west chute of the island was closed, to secure the landing upon the opposite shore, and six spur dams were built, one of which was removed later. Two miles and a half below this island, a closing dam at Denmark Island forms a part of the system. Altogether upon these works, were used, in construction and repairs, 90,420 cubic yards of stone and brush, besides large quantities of gravel, at an expense of $151,486.73, including all expenditures.

It may be added here, that the works seem to have been perfectly successful, and reasonably permanent; and navigation has been uninterrupted since, at these points. At Hickory Chute, below Denmark Island, and above Louisiana bridge, are other works of construction. From Saverton to Louisiana bridge, 25 miles, the total expenditures were $198,703.78.

In the next stretch, from Louisiana bridge to Hamburg, 25 miles, was found the extensive deterioration in the river bed, due to the Louisiana railroad bridge, which was so built as to cause extreme contraction at high water, and consequently extensive scouring, and bars lower down, as a necessary sequence. To redeem the channel, low water contraction works were made, and the threatened banks protected, at large cost. In these works 21,872 cubic yards of stone, and 17,758 cubic yards of brush, were used, at a cost of $50,988.77.

At Slim Island, four closing dams, and one spur dam, were built; consuming, with the necessary shore protection, 37,228 cubic yards of material, at a cost of $37,963.09. The total expenditures on this stretch of river were $96,788.04.

The lowest stretch, from Hamburg to the mouth of the Illinois River, 48 miles in length, received attention at Westport Chute, Bolter's and Dardenne Islands, at a total expense of $41,968.30; principally spent at the last named locality.

In surveys and gauges, snag boats, and wrecking, the procurement of supplies,— afterwards used,— and of steamer and barges, were expended $134,940.31. The total account showing an expenditure of $602,406.76; and the total appropriations for the same time $615,000.00. The act of 1884 gave $200,000 additional, which is being expended in continuation of the works below Louisiana bridge, works at LaGrange, Quincy Bay, Bolter's Island, and in repairs.

In addition to the appropriations above quoted, special sums have been granted for the following harbors, within this portion of the river: Alexandria, $16,000; Quincy, $45,000; Quincy Bay, $25,000; Hannibal, $50,000; and Louisiana, $10,000. In construction, works

7

of the character described, have been built; and at three points, more or less dredging, besides, has been done. The landings at Hannibal and Louisiana have been bettered, and the winter harbor of Quincy Bay.

In making a short review of the commerce of the river, it is intended to present such facts (of which the reports are, fortunately, full and exact), as will show the business at different points, and the character of the business. But in drawing deductions from these figures, it is well to remember that the large increase in commerce and manufactures, which culminated in the United States in 1881, and is already noted in connection with other statistics, elsewhere, finds its influence exerted upon this river, as is plainly seen, if refer-ence is made to the fact stated previously, that the low water season of 1879, was marked; yet, this would not so appear from the statistics.

STATEMENT OF STEAMERS PASSING VARIOUS BRIDGES.

BRIDGES.	Miles from St. Paul.	1875.	1876.	1877.	1878.	1879.	1880.	1881.	1882.	1883.
Hastings.............	27	.....	.....	.....	.....	2,468	2,065	1,843	2,563	746
Winona.............	125	.....	.....	.....	2,948	3,760	4,593	4,739	5,627	4,893
Dubuque	284	2,771	2,647	1,801	2,189	2,894	3,625	3,865	4,345	3,585
Rock Island.........	383	1,830	1,976	1,560	1,643	2,514	2,664	2,711	2,593	2,561
Burlington..........	467	1,412	1,820	1,139	1,318	1,251	1,871	2,312	2,323	1,943
Hannibal...........	563	1,370	1,863	1,467	1,393	1,586	1,925	1,909	3,031	2,886

STATEMENT OF RAFTS PASSING SAME BRIDGES.

Hastings.............	27	.....	.....	.....	.....	54	20	14	60	.....
Winona	125	.....	.....	.....	868	1,889	1,392	1,777	1,939	1,352
Dubuque	284	1,136	1,090	642	694	801	1,375	1,422	1,022	1,380
Rock Island.........	383	618	627	413	352	571	782	953	984	972
Burlington..........	467	*164	*209	*155	254	367	522	523	425	379
Hannibal	563	168	183	181	156	260	237	322	300	262

* Partial record.

FEET, BOARD MEASURE; THOUSANDS.

	1875. M.	1876. M.	1877. M.	1878. M.	1879. M.	1880. M.
Lumber manufactured on the upper Mississippi River................ }	1,075	1,178	860	826	1,274	1,732
Of this received at St. Louis...	.....	166	163	129	180	230

Tons of freight received at Saint Louis from the upper Mississippi:

1871................236,887		1878.................174,065
1872................242,584		1879.................221,285
1873................281,175		1880.................226,095
1874................231,060		1881.................190,815
1875................198,100		1882.................135,540
1876................224,860		1883.................126,330
1877................136,715		

Of this freight some further particulars are given.

RECEIPTS AT SAINT LOUIS FROM UPPER MISSISSIPPI RIVER.

ARTICLES.	1875.	1879.	1880.	1881.	1882.	1883.
Wheat, bush.	748,968	....	761,323	444,686	398,498	335,443
Flour, bar'ls	91,529	....	53,958	52,137	79,828	65,876
Corn, bush...	832,486	535,802	959,717	698,630	372,560	82,675
Oats, bush...	1,050,440	945,557	1,594,169	1,697,834	1,114,178	811,062
Hogs, No....	35,681	32,182	35,824	....	....	....

STATEMENT SHOWING COMMERCE OF THE DES MOINES RAPIDS CANAL.

Fiscal Year.	Steamboats. No.	Barges and flats No.	CARGO.			RAFTS.		Lock'g's at one lock. No.
			Passengers. No.	Gen'l M'dse. Tons.	Grain. Bushels.	Logs & Lumber B M- M	Shingles & Lath. No.—M.	
*1878..	670	548	....	53,346	737,415	25,000	7,700	824
1879..	802	454	5,008	64,658	2,192,642	41,434	20,471	1,564
1880..	967	651	13,231	78,989	2,197,469	44,993	58,425	2,497
1881..	840	276	10,003	44,962	1,154,092	63,270	26,749	1,339
†1882..	760	444	8,588	29,043	781,817	21,625	8,008	2,292
†1883..	1,107	705	9,192	43,359	729,174	14,133	15,993	1,353
1884..	913	245	13,057	54,215	470,500	64,418	41,107	1,908

*Opened August 22d, 1877; closed 25 days for repairs.
†Good stage water; river used more than usual.

An examination of these statistics will show that while the upper Mississippi is not a great through route in the sense of the transportation of large amounts of freight throughout its distance; yet upon every portion of the river large transfers of freight of all kinds and especially of lumber, annually are made; that the traffic is well distributed through the whole, in quality and quantity; that grain and flour do not follow the river in large quantities, and these amounts are not increasing.

It will be remembered that throughout its whole extent, the river has one, and through a large part, two railroads along the shore; if it be also remembered that the river is closed during from four to five months in winter, the conclusion will be that as a local factor in transportation it has importance. The lumber and rafting business is now of great importance, and probably will diminish before long; but on the other hand, the reclamation of large tracts of fine bottom land from overflow, by levees now in progress, will in time change forest and swamp lands to farms, with consequent increase of river business.

In the large area, and consequent full supply of water, the Upper Mississippi is fortunate. The gentle and fairly uniform slope, not only of the river itself, but of all of its tributaries, has secured a moderate range between high and low water stages, and the easy transition from the one to the other condition. The comparative freedom from sediment transported, and the general similarity in character of the river bed, and its accompaniments, throughout the course of so many miles, and the gradual increase in difficulty of rectification and control, all combine to make the problem of improvement one of great interest. Now that the permanent work at the two great obstructions has been done, and enough of the general plan of improvement carried out to warrant an expression in favor of its success, it will be in the future to determine if the extension down stream meets with as uniform success as has been encountered in the upper portions, and at the worst places. It will be finally decisive of the great question, whether rivers can be permanently improved by dams and dikes; if a full trial is had upon this, the most favorably characterized river in this country upon which to attempt the plan. But temporary improvement at some points, or closure of secondary channels alone, will not be decisive; nor will the securing of a light draught channel in favorable seasons over an improved portion, provided impassable obstructions above and below remain, particularly if they were not formerly in existence at those places.

CHAPTER X.

RESERVOIRS AT THE HEADWATERS OF THE MISSISSIPPI RIVER, FALLS OF SAINT ANTHONY, AND UPPER TRIBUTARIES.

The subjects of this chapter have each their own interest, from an engineering, or commercial, point of view, and are grouped separately to preserve the unity of treatment of the preceding chapter.

It is difficult to treat them concisely, and yet, the space allotted must be short.

RESERVOIRS AT THE HEADWATERS OF THE MISSISSIPPI RIVER.

It has been a favorite idea, that the radical improvement of certain streams could be effected by the impounding and retaining of water, which, falling upon the areas near the sources, would, if not so retained, pass away in great floods. By afterwards allowing this surplus to flow during the low water season, an equilibrium would be established throughout the year, which would be beneficial during both seasons. Hardly any elaborate scheme of river improvement has been discussed without reference to this idea of reservoirs. Upon the Ohio, Mr. Ellet proposed, and elaborated the system, in theory, upon a large scale; but no actual construction was ever undertaken upon the grand scale, now in progress upon the upper Mississippi; and its results must be watched with interest, from an engineering, as well as com-

mercial, point of view. [1] Beginning with 1869, Major Warren asked
for $10,000 to make surveys, and reconnoissances, above the Falls of
Saint Anthony, with the view of making large reservoirs on the headwaters of the Mississippi, to aid in keeping up the navigation at low
stages. He adds "whether practicable or not, the question has
attracted a great deal of attention." [2] From the surveys so made,
locations were suggested for three dams, to form reservoirs, with a
total capacity of fifty-four millions of cubic feet. The subject was
not at all elaborated, however, until it came up in 1874, as the second
subdivision of the Mississippi Transportation Route, when Major
Farquhar made a report, [3] based upon an extended survey for these
sites, which gave the levels, transit lines, meanders of the lakes and
rivers, and careful gaugings.

The report gives a careful description of the country, and of the
proposed sites, the areas of water shed being 19,903 square miles;
and lastly, of the amount of water to be retained.

Sites for eight dams were selected, and estimates made for their
construction, with full particulars. This report was of course fuller
and more specific than the preceding. The subject was again brought
up, by the action of Congress in 1878, calling for special surveys, and
allotting $20,000 in 1878, and $25,000 in 1879. Capt. C. J. Allen
was placed in charge of the investigation, and has remained in charge
of the work. His first, and very full report, [4] covers the general
subject as far as the headwaters of the Mississippi itself, and gives a
vast amount of details. Certain of the elements are easily determined.
The drainage areas of the various streams; the levels and contours of
the reservoirs, and the cost of construction of the dams. The amount
of rainfall, at points within the region under discussion, and the
discharges of the different streams, at different times, are also easily
ascertained. But the remaining elements are not so apparent. The
relations between rainfall, and the discharge of small streams, and
accumulation in reservoirs; and the relations between allowed outflow from reservoirs and the discharge of the river at some distance
below them; and the influences of evaporation; and absorption of
soils, are all complex and extensive.

The literature upon the subject is of necessity extensive, and these
reports could not be condensed if desirable. But it will suffice to
show the care with which the whole matter has been considered if it
be noted that the reports would make a fair sized volume themselves,
covering 251 octavo pages, with numerous maps and drawings. Besides these features, which called for a satisfactory answer to the problem
before construction could begin, there were many others arising after
operations began, which do not ordinarily come up in river work.
The building of dams would cause overflow, and the rights of private
parties, and of Indians, could not be preserved without recourse to

1. 1869, p. 189.   2. 1870, p. 284.
3. 1875, II, p. 434.   4. 1879, II, p. 1193.

the legislatures of the two states of Minnesota and Wisconsin and the action of Congress. In this brief outline we have only the sketch of the work involved. When the sites were selected, and a plan of construction drawn up, the whole engineering features came to be discussed and adopted or modified by a board of four engineer officers, so that considerations of every kind arise in succession.

Capt. Allen selected sites for seven dams, which are given in the following table, and estimating the cost of this series at $336,458.60, recommended that an experimental dam be built at Lake Winnibigoshish. The cost of maintenance and repairs for ten years after completion was placed at $10,000 per annum, and $7,840 per annum for the cost of operating:

RESERVOIRS ON THE UPPER MISSISSIPPI.

LOCALITIES.	Height above stage of 1874. FEET.	Length of dam. FEET.	AREAS OF		Capacity of Reservoir. cub. feet. millions.	Supply of water. cub. feet. millions.
			Basin. sq. miles.	Reserv'r. sq. miles.		
Winnibigoshish............	14	1,114	1892	154.7	45,754	37,774
Leech Lake................	4	3,300	1001	218.5	22,568	15,461
Mud Lake.................	6	1,000	160	17.3	2,885	3,138
Vermillion...............	10	2,300	433	34.5	5,771	8,563
Pokegamma..............	7	400	179	23.6	3,752	5,118
Total Area and Capacity	....	....	3665	448.6	80,730	70,054
Gull Lake.................	12	450	272	260.	6,134	5,266
Pine River................	12	600	551	....	4,913	10,667

The examinations of the Saint Croix and Chippewa Rivers begun that year, were not finished and reported upon until the following year, when these and the Wisconsin River were discussed. As no construction has as yet begun upon these streams, they will not be here included. [5] But the general conclusion was drawn that if the whole system as worked out was considered "collecting the various items of discharge, we find that we can control from the proposed reservoirs at the sources of the Mississippi sufficient water to insure a steady flow of at least 12,200 cubic feet per second past Saint Paul for 100 days, and from the Saint Croix 6,000, for the same period."

Congress having appropriated $75,000 on June 14th, 1880, for an experimental dam at Lake Winnibigoshish, a Board of Engineers was called and reported August 5th, 1880, modifying some of the details of the plan proposed by Captain Allen. But owing to delays caused by the necessary legal and legislative action required, (as alluded to), actual work did not begin till the following year. [6] Mean-

---

5. 1880, II, p. 1611. 6. 1881, II, p. 1761.

time the surveys were extended to include Rock River, Illinois and Wisconsin, and other similar work was completed. Revised estimates for the seven dams proposed in the table above, amounted to $538,135.80, and including necessary telegraph lines to $573,660.08. The study of rainfall, drainage, absorption and evaporation was continued, and large material accumulated.

[7] In addition to the meteorological and hydrometrical observations, the three gauging stations; on the Mississippi River above the Crow Wing; on the Crow Wing; and on the Saint Croix River; were continued for an entire year, ending with December, 1882. The three areas thus gauged are 7,283 square miles; 3,576 and 5,950 square miles in the order above. From daily observations of these gauges, and from the rainfall upon these areas, the actual available precipitation for each area was determined, and the ratio to rainfall. From these actual, not theoretical data, and the average rainfall, it was deduced that the reservoirs for the Mississippi above Saint Paul, would furnish a quantity of water at Saint Paul equivalent to 5,500 to 6,000 cubic feet per second for a period of 90 days; this covering the time of low water and giving an amount enough to render unnecessary all wing dams above Minneapolis, and to double the present lowest water discharge at Saint Paul.

The report of 1881 gives Capt. Allen's proposed dam, in details, together with the modifications proposed by the board.

From the fact that the work lay at a distance from sources of supply and of labor, it was necessary first to work the roads leading to the site, and an attempt to let the items by contract developed the fact that labor could be procured more economically by the government than by accepting the bids. Machinery, supplies and tools were procured, quarters and warehouses were built, and arrangements were made with the Indians, in reference to the overflow of their lands.

Work upon this dam has progressed rapidly, and the reports for 1884 stated that it would be completed during the fall of that year. Meantime Congress appropriated $150,000 in 1881; $300,000 in 1882; and $60,000 in 1884, in continuation of the project. As soon as sufficient funds were available, work began on the second great reservoir, that of Leech Lake, in the fall of 1882, and supplies and machinery were bought, and the lumber for the dam at Pokegama Falls. During that year the board of engineers on the Winnibigoshish dam was again called together, and considered and made certain modifications to the dam, as first adopted. [8] The dams begun in 1882, also are reported as finished during the fall of 1884, work having been pushed, winter and summer, during the prior year. [9] Preliminary work for the Pine River dam began in the winter of 1883-4 in procuring about half the timber and piling necessary, and in the detailed surveys. This structure is to have a lift of 17½ feet, built of timber crib work, the upper 5½ feet will be obtained by the use of stop plank above the

7. 1883, II, p. 1457.   8. 1883, II., p. 1471.   9. 1884, III., p. 1614.

wasteway. The total expended to July 1, 1884, was $418,367.38. In order that navigation may be benefitted upon and below the Saint Croix River, upon the Chippewa, and the navigable reaches of the Wisconsin, the system of dams proposed for each must be carried out; and no benefit of consequence to the Mississippi below Lake Pepin can be predicated, unless the entire system be built.

Logging operations in these regions have been embarrassing during construction, and the attention of Congress invited to the fact, and that some action is desired to protect the reservoirs for the interest of river navigation.

Two reservoirs were sufficiently advanced in the fall of 1884 to control navigation on 160 miles of river where there are three steamers. By October three reservoir dams were completed. From April 20th to June 1st, 1885, they kept up navigation upon this piece of river; after June 1st the discharge was increased to from two to three times the ordinary low water discharge. The figures of 1883 are borne out by the facts. The fall of 1885 will probably demonstrate the working of the completed reservoirs, to some extent. It should be borne in mind that the advocates of the system have never claimed more than that reservoirs might be a desirable adjunct to river improvement systems.

PRESERVATION OF THE FALLS OF ST. ANTHONY.

[10] In a letter dated January 9th, 1869, written to the Chief of Engineers by Maj. G. K. Warren, the writer calls attention to the recession of the Falls of St. Anthony, owing to various causes, and uses the following language: "The stoppage of this recession is a work of great difficulty and expense, and as it concerns the future navigation of the river, as well as the existing water power, it is a question for public consideration."

[11] This language was adopted by the Chief of Engineers, and an appropriation made July 11th, 1870, of $50,000. Under these circumstances it might not be proper for me to question the statement as to the influence upon the navigation of the river, which the preservation of the falls has had. [12] No steamers have navigated the river in the vicinity of the falls for seven or eight years past. The original report upon the threatened danger closed with the remark "besides affording inestimable protection to manufacturers and those interested in the mill and water-power companies, who have expended a large amount of capital in improving the water-power afforded by the Mississippi River passing over the Falls of St. Anthony, and have built the two cities at the Falls, and thus have added greatly to the wealth and production of the State." But however this may be, the operations themselves have been of great interest in an engineering point of view, and will be glanced at briefly.

---

10. 1869, p. 211.   11. 1870, p. 58.   12. 1884, III., p. 1597.

[13] Col. Macomb, succeeding Maj. Warren, in charge, proposed to close a tunnel built under Hennepin Island by private parties, which had been breached by the river, and threatened a general erosion and recession not anticipated when the first danger was reported. A coffer dam was built, beginning in 1870, and continued the following year, around the breaches into the tunnel, of 1,268 lineal feet, intended to sustain a head of water 12 to 16 feet. A wall was then to be built around the upper breach, and the gorge was to be filled with clay and gravel. [14] The usual chapter of accidents resulting from breaks and leaks could be written as to the summer's work, but by November 23d, 1871, the wall was completed. Meantime a new break had occurred from the river into another portion of the tunnel, and new coffer dam, 755 feet long, was required, and more wall, and puddle—and money. August 10th, 1872, a Board of Engineers considered the subject, and approved of existing plans, confining work to the preservation and strengthening of the coffer dams and walls just made, and the ultimate lining and plugging of the tunnel, the cause of the mischief. The private interests of Minneapolis, so highly jeopardized by the threatened loss of their water-power, had joined in with means and advice, and all plans of operation harmonized.

The record of 1873 contains only an account of the unfortuate giving way of a coffer dam upon April 15th, with consequent large destruction; and on May 17th a great breach under the earth dam at the head of the ledge; and the repairs necessary to replace matters in as good a condition as before the accidents. However, the whole was finally and successfully done, the breaks repaired, and the tunnel filled. By this time $200,000 had been spent, besides the amount subscribed by private parties, in bringing back a state of affairs as good as when operations were first proposed.

[15] A board of engineers, composed of Colonel Macomb, Lieutenant-Colonel Kurtz, and Majors Weitzel, Poe, and Farquhar, (the last named, now in charge of the work), met April 15th, 1874, and recommended the building of two crib dams, bolted to the rock and filled with stone; the repair and extension of the wooden apron, begun by private parties, to prevent undermining of the falls; and lastly the building of a dike of concrete across the river, along a line a short distance above the apron on the Minneapolis side. This dike was to be constructed in a tunnel to be excavated on the proposed line, and resting upon the homogeneous stratum below the soft stratum, whose destruction caused the trouble. The project proposed was estimated at $420,000.

The construction of this concrete dike was begun in July, 1874, and finally successfully completed in November, 1876. In conception, and in execution, the plan was novel, and difficult; and its engineering details would be of much interest if repeated. The apron was repaired, and the rolling dams proposed were completed by 1878;

---

13. 1871, p. 295.   14. 1872, p. 297.   15. 1874, I, p. 283.

and since that time, work has been done mainly to preserve the status, and repair damages caused by ice, and flood, and obstructive action of private parties. In 1883, (Major Allen being the officer in charge) the repairs of crib work called for two new cribs, one 67 feet long, 26 feet wide, and 28 to 33 feet high; the other 8 feet long, 20 feet wide, and 26 feet high; both cribs being sunk under the toe of the apron. This, and similar repair work, consumed more than a quarter of a million feet, board measure. A crib was built, and sunk below the falls, to prevent further scour of the river in the angle, and also to act as a buttress to strengthen the crib work under the toe of the apron. This crib was 80 feet square, and 6 feet high; built of 12-inch square pine timbers, solid side walls, and eight open interior cross-walls, filled compactly with stone, and grouted. It is decked with 12-inch square timbers, drift-bolted to each other, and the bottom and deck of the crib are tied together by vertical iron rods.

MATERIALS USED.

Pine timber, feet, board measure	194,972
Iron, pounds	43,869
Stone, cubic yards	827
Cement, barrels	353

Previous to the sinking of the crib, the bed of the river had been leveled up with stone, and after it was sunk it was rip-rapped with stone, of which 3,023 cubic yards, in all, were used.

The total amount of appropriations made for the Falls of Saint Anthony has been, in twelve items, $615,000; which should not be considered as an item of the improvement of the upper Misssissippi, of which it forms no part.

Above the Falls of Saint Anthony, appropriations have been made since 1874, to 1882, amounting to $80,000, which have been expended in the removal of obstructions. The season of 1883, was of unusually low water, and without this work, steamers would have been compelled to lay up. There is a general depth in the channels now, of three feet; and, as already noted, the reservoirs will undoubtedly furnish a continuous boating stage.

Three steamers were in use, in 1884, above Grand Rapids.' Ninety millions of feet of lumber were floated in 1883. The logs run above Saint Cloud amounted to three hundred and thirty millions of feet.

Upon the Minnesota River, from 1867 to 1878, were expended $117,500 in removing obstructions. Commerce existed at one time upon the river; but none is found there now, and appropriations have ceased.

The Chippewa River is the great lumbering river of this region; in 1883 four hundred and fifty million feet, board measure, in logs, and two hundred and sixty-nine millions in manufactured lumber, came down the river. Since 1876, $98,000 have been spent, very judiciously, in works of construction, which have increased the navigable depth enough to allow three steamers, in 1883, to make use of the river, besides facilitating the immense lumber business.

At Yellow Banks, on the Chippewa, are extensive high sand bluffs or banks, which wear away and supply masses of sand to the Chippewa and the Mississippi. An appropriation of $30,000 in 1882 was used for driving rows of piles at the foot of the slopes, and then by the use of brush fascines, a revetment of brush was carried to 10 or 12 feet above low water mark. When growing willows are established the wearing away will not be continued. About 23,000 linear feet of bank were to be protected. Thus far 4,969 linear feet are entirely, and 3,145 partly complete.

Upon the Saint Croix was expended $75,000 from 1878 to 1884 in removing obstructions. Three steamers made 141 round trips in 1883; and 270 millions feet, board measure, in logs, and 167 thousands in manufactured lumber, came out that year.

WISCONSIN RIVER.

[16] The Fox and Wisconsin Rivers formed a route for the early discoverers, and subsequently a line for military operations, and a highway for trading and emigration. Aided by a grant of land from congress in 1846, the legislature of Wisconsin undertook to connect the navigation of the two rivers by a canal. This enterprise was turned over to an incorporated company in 1853. In 1866, this company having failed to perform fully its agreement with the State, the trustees sold at public auction the improvements and franchises, and the purchasers organized as a new company. Congress directed in 1866, a survey to be made of the route, and Major Warren was assigned the duty. His elaborate report was submitted in 1875, printed in the Report for 1876, 2d part, pages 189 to 298, with numerous maps and plans. By act of 1870 the Secretary of war was authorized to adopt for the improvement of the Wisconsin "such plan as may be approved by the Chief of Engineers," and to negotiate with the Canal Company for a transfer to the United States of its franchises and property, upon and between the two rivers.

An agreement was reached and the transfer made in October, 1872, at a cost of $145,000; the company retaining its personal property and the water powers created.

The two rivers had been treated differently. The great fall of 170 feet, from Lake Winnebago to Green Bay in the lower Fox, through a distance of only 37½ miles, rendered it necessary to apply there the system of locks and dams, which was also applied to the upper Fox, though the slope here is much less. A proper depth can be obtained in this way through the whole length of the Fox. The case of the Wisconsin is different. The length of 118 miles, with a total fall of

16. 1884, III, p. 1900,

177 feet, gives an average of 18 inches to the mile. This slope taken in connection with the shifting nature of its bed, composed of sand, and the small discharge, led General Warren to reach a conclusion adverse to permanent improvement of the river by wing dams, or by locks and dams, and to advocate instead, a canal along its banks, with 4-feet draft, and locks 165x35 feet in the clear, at an estimated cost of $4,000,000 ' to be built in two years, with the full amount appropriated.

The improvement, however, of the bed, by means of contraction of water way with wing dams, had been previously put on trial, but was not sufficiently extended to furnish conclusive results. The officers in charge, Colonel Macomb, and subsequently Major Houston, did not agree with Major Warren's conclusions. Operations were carried on with varying success, and indefinite results and opinions, until 1879, when it was thought best to suspend operations and await a report from a Board of Engineers, to which the plan of improvement had been referred.[17]

At this time there had been built in all 152 dams, with a total length of 68,489 feet, and 8,189 linear feet of bank protection—over stretches of river from Portage City, 24 miles, to a point below Merrimac, and from Lone Rock to Boscobel, 30 miles. At this time the board directed further work to be done upon 12 miles of the Portage section, and certain observations to be taken, before deciding upon results. Upon January 5th, 1884, this board reported, in compliance with instructions dated May 22d, 1883, upon the subject, in a very full and minute report. [18] Discussing the conditions of the river at different periods beginning with 1867, and examining the gauge readings, discharges, channel depths, and duration of stages, to November 30th, 1882, the board decided that while final results were not as yet attained, it appeared that the mean depth of 2.46 feet, and least depth of 1.1 feet of the year 1867, had been increased to 3.27 feet and 2.88 feet respectively in 1882. Assuming that a depth of at least 4 feet was necessary for a water route of through transportation from west to east, the board then proceeded to inquire into the conditions necessary for its attainment upon the Wisconsin. The opinion was expressed that in order to secure this the width of the river at extreme low water should be at Portage, 140 feet, and at Bridgeport, 300 feet; that the increased velocity caused by this contraction would be an evil; that the use of wing dams without training walls, on each side of the channel in line with their ends, would lead to great inequalities in depth and the formation of shoals; and finally that to complete the improvement as begun, with the modifications called for, would require 538,964 lineal feet of dams, in addition to those built, at an estimated cost of $2,155,856, and adding 830,720 lineal feet of training walls at a cost of $3,322,880, the total cost would be $5,478,736 for the work.

---

17. 1879, II, 1533.   18. 1884, III., p. 1900.

The board also thought that the estimate of General Warren, for an independent lateral canal, should be perhaps doubled. The study of the Wisconsin River is especially recommended to those who believe that a radical improvement of a river with a high slope and shifting bed, can be effected by contraction methods.

The amount expended upon the Wisconsin to June 30th, 1883, was $570,117.40.

CHAPTER XI.

MISSISSIPPI RIVER BETWEEN THE ILLINOIS AND THE OHIO.

This portion of the Mississippi has been specified in the River and Harbor Bills, for many years past, as a district in itself. Between the Illinois and the Missouri, there is no difference in the character of the river and of work done upon it, from the Upper Mississippi; and it will not be necessary to devote time to this stretch of twenty-four miles, further than to state that a closing dam was built from Piasa Island, 7 miles above Alton, to the Missouri shore, and subsequently removed in part; and that a closing dam was built across Alton slough, and subsequently a long training dike from the Missouri shore down to Alton Island, both works being intended to remove a large sand bar which obstructed the river front at Alton. This bar has disappeared under the influence of the contraction, and both the effect and the constructions are believed to be permanent.

[1]From the mouth of the Missouri to Commerce, a distance of one hundred and sixty-two miles, is a section deriving its distinguishing features from the Missouri River, but is distinct from the alluvial region found below Commerce. Turbid waters, shifting bars and channels, rapid erosions of alluvial banks, and extensive accretions, building up and removing islands, tow-heads and battures with great rapidity are characteristics. At the higher stages, the crumbling banks falling in masses, the spoil of the forests covering the surface, and the boiling, swirling current show the power of the stream. At low water the wide wastes of sand bars, bristling with snags, and drifts of every shape and size, are alike suggestive.

From the mouth of the Missouri to St. Louis, 15 miles, the river does not touch the bluff. Excepting a rock formation at one point, there is nothing to check erosion on either side. Below Saint Louis the river follows the Missouri bluff closely for 55 miles, the only exception being at Rush Tower Bend, where a former island has become connected with the Missouri shore. Below Saint Mary's and Cape Cinq Hommes, the river is winding to and from the bluffs. At the last named place the river is at its narrowest, and rock appears on both sides, and just below here, at Liverpool, the main Illinois bluff recedes from the river, which continues along the Missouri bluff, remaining near it as far as Cape Girardeau, where the main bluff leaves the river and appears no more. A short distance below Cape Girar-

1. 1875, II, p. 472.

deau a depression allows the Mississippi waters in floods to escape into the swamps, and thence into the Saint Francis. Isolated or detached bluffs are on either side from Cape Lacroix to Commerce, where the valley expands into the great alluvial basin of the Lower Mississippi. Throughout this section the river is, as a rule, held on one side by rock bluffs, and is remarkably direct in its general course, only when it leaves the bluffs does it work out long sweeping curves. Two tributaries of considerable size, the Meramec and Kaskaskia enter, but their effect upon the river is slight.

The valley is from three to eight miles in width except near Grand Tower and the Grand Chain, and is subject, in great part, to overflow in time of floods. The ground generally slopes back from the river to the sloughs and lagoons, with which the bottom is interspersed, and in like manner from the the farther bank of the slough or lagoon. From Commerce to the mouth of the Ohio, thirty-seven and a half miles, the alluvial region is traversed, and gives its distinguishing characteristics to the section, the uniform texture of the soil allowing the river to shape its course without restriction.

The times of floods of the Ohio and Mississippi are very different, and at high water upon the Ohio, its back water influence frequently extends farther than Commerce. When the Mississippi is high and the Ohio not, the conditions are reversed and the current becomes very rapid. Owing, in great measure, to these excessive changes of velocity, the channel is very unstable, and the erosion and accretions extensive.

From the Missouri to the Ohio, the bed of the river is so broad that the channel meanders from side to side within the bed, just as the bed itself meanders within the valley from bluff to bluff. The movement of the bed is ordinarily slow, but the shifting of the channel is continual and often in sudden leaps, forsaking one course and cutting out a new one in a very different direction with very little warning. From discharge observations made in 1880-1881 at Saint Louis by the Mississippi River Commission, the following means are quoted of area of cross section, velocity of current and discharge:

GAUGE. feet.	MEAN VELOCITY. feet per sec.	AREA CROSS SECTION. square feet.	DISCHARGE. cub. ft. per sec.
24	3.08	32,000	97,000
26	3.40	36,000	125,000
28	3.70	40,000	155,000
30	4.05	45,000	188,000
32	4.38	50,000	222,000
34	4.70	55,500	265,000
36	5.04	61,800	306,000
38	5.35	68,000	358,000
40	5.68	75,000	413,000
42	6.03	82,500	480,000
44	6.35	90,000	552,000
46	6.66	98,800	640,000
48		107,500	742,000
50		117,000	865,000

The lowest gauge reading and discharge there recorded were on January 27th, 1881 ; gauge reading, 22.74 ; area, 20,609 square feet ; velocity, 2.20 feet per second ; and discharge, 45,330 cubic feet per second. It was calculated in 1875 from observed discharges taken in 1873 and 1874, that the probable low water area would be about 15,000 square feet, and discharge about 36,565 cubic feet per second ; but extreme low water discharge had not been measured.

The maximum of 336 parts in 100,000 was noted in 1881, July 2d ; in the sediment observations of that year, the gauge read 39.41, and the river rising on that day.

The unstable character of the Mississippi arises from the rapidity of its currents, the excessive variations of volume, and the loose texture of the soil through which the river works its way. Soundings have shown that, in general, the depth of water in the river does not follow the rise and fall as given by the gauge readings, and the rise of water is followed by a wave of sand, moving at a slower rate. The depth upon a bar, at various low stages, often increases as the river falls, and diminishes as the river rises. In the language of boatmen, the bars "cut out" in a falling, and "flatten out" in a rising river. The bars, composed chiefly of movable sand, travel down stream at a rate in proportion to the velocity of the current, changing in shape as they pass the bends in the river, or meet with obstructions, that lessen the velocity or deflect the current. When the river rises this movement of the bars is more rapid ; and as the water recedes the channel is cut through a crest of the bar, generally at its lowest point, which is usually nearest the shore, leaving a pool of water below each bar and a channel winding from side to side.

The great variations found in width, either at high or low stages, make it impossible to give a summary of the natural width of the river, without having recourse to too much detail. Low water widths from 1,730' to 3,740' were noted at different localities in 1873 and 1874, the gauge reading from 6'.9 to 11'.75 above low water. Colonel Simpson's report, which has formed the basis of this description, estimated 3,500 feet as a suitable width for a bank-full river, 25 feet above low water. The Saint Louis observations gave a width from 1,260 feet to 3,347 feet between low and high water.

No operations, other than for the removal of snags, were had until after the passage of the act of 1872, which gave $125,000, and of 1873, $200,000, for this district. The closing of Alton Slough, already alluded to, was the first work ; and the protection of Venice and Sawyer's Bends, near Saint Louis ; and the general regulation of the river in this vicinity, as planned by a Board of Engineers, consisting of Lieutenant-Colonels Newton and Raynolds, Majors Warren and Merrill, and Captain Allen.[2] All of the dikes and dams, and shore protections, first constructed, were of the combination of brush and

2. 1872, p. 358.

stone, found most economical and advantageous upon sandy or moderately secure foundations, which will support broad, and low, weighty structures. Later on, similar regulation was attempted at the first great obstruction encountered below Saint Louis, at Horsetail Bar.

[3] The term Horsetail Bar includes a stretch of river about five miles long, extending from the southern boundary of Saint Louis to the foot of Carroll's Island. The average width, when the works began in 1873, was about 5,000 feet. There were several large movable bars which obstructed the navigation, and it was no uncommon thing to find as little as 4 feet depth of water in the channel during the low stages of the autumn, the obstacle being sometimes at one point and sometimes at another. The bottom was composed of shifting sands and mud. It having been decided to undertake the contraction of the river to a width of 2,500 feet, a series of dikes perpendicular to the shore was begun. The outer extremities were to be connected by longitudinal dikes, or training walls, located upon the lines of the new banks. The construction of one of these training walls was begun in 1877. The object of these works was to confine the water of the river within the prescribed limits by means of their own solidity and weight. They were built of rip-rap upon a foundation of brush. The great volume and velocity of the Mississippi; the treacherous nature of the foundations of the dikes; their necessarily great length, and the intervals between them ; and lastly, the destructive efforts of the ice; all conspired to attack, and finally overcome these works; which were breached and destroyed in part, or in whole. [4]To support the training wall, which originally had a length of 5,950 teet, it was determined in the spring of 1879, after an extensive breach had been made in it, to connect this training wall with the shore by a series of continuous hurdles, built at intervals of about 100 yards. A row of piles, 5 feet apart, was driven, and between them courses of willow brush were interwoven. After the wattling had been completed pieces of brush were pushed into the open spaces until filled. The lowest courses were pushed as far as possible below the surface, (which proved to be to a distance of 8 feet), and if the depth of the water were greater than that, a free space was left of necessity between the edge of the hurdle and the bottom. The object of these hurdles, which was to secure a deposit of sediment from the river, was obtained at once and the success was marked. The training wall backed by a mass of accretion from the river was made more secure than by any other known support possible to be procured. The success of this work led to its extension, and the brush and stone jetties were abandoned. [5]At many places the water was too deep for the advantageous use of piles. In such cases brush obstacles were constructed either continuous or in sections upon floating ways from which they were launched. One side was secured to the bottom by anchors, and the other supported by buoys. Many details were devised and attempted in this system of slight and temporary obstacles.

---

3. 1880, II, p. 1362.   4. 1879, II, p. 1028.   5. 1880, II, p. 1362.

Hurdles were built to a height of 15 feet above low water, and additional height desired was to be secured by a growth of willows over the reclaimed area.

It will be remembered that at the time similar constructions were in use upon the Missouri River. A mattress designed to protect the bottom from scour, built in continuous length, and deposited in position from the boat where made, was combined with a permeable suspended component to check the current, and secure deposits .in the rear of it. The loss in 1880 of a long continuous mattress with curtains attached, when being placed in deep water, with a strong current crossing the line of the wall, led to a modification of design. Many details of construction and plates of illustrations are given in the report for 1880. By 1881 it was decided that for all depths less than 35 feet the most efficient support for the brush obstacle was the pile, which acted as both anchor and buoy ; and all piles were braced, except in depths of 6 feet or less.

By the year 1881, had been built at Horsetail 42,630 lineal feet of hurdles, and within the area of their action had been deposited by the river, within two years, over 13,000,000 cubic yards. The change in direction of the channel was the first advantage gained, and subsequently an increase of depth ; though the first action was attended with a decrease in depth.

Although at this time work was carried on at various points within the district, it was decided and approved by superior authorities in 1881 to make the improvement continuous by beginning work at Saint Louis, and carrying it on down stream, where needed, each work to bear intimate relation with that done above it. During the low water season of August, 1881, the middle reef at Horsetail assumed a troublesome form, the Illinois end swinging down stream, so as to give it a length of nearly two miles. There were two depressions or channels through it, the depth, in that used by steamboats, varying, but reaching at one time a least depth of six feet. At this time it was thought best to aid the river, and, by the use of the water jet, the channel depth was increased $2\frac{1}{2}$ feet with ten hours' work. A least depth was formed that season, upon the lower bars, of 5' at Cairo Point, and $5\frac{1}{2}'$ at Jones, and Eliza Point. The officer in charge, Maj. Ernst, thought, therefore, that the result sought for at Horsetail had been secured, and that if the existing works could be maintained, and the new deposits protected upon the top and sides, by the growth of willows, a sufficient channel would be found at all seasons.

The next year, 1883, found that the area upon which willows were growing had been much enlarged. The building of one new hurdle, 2,450 feet long was required, and repairs to old work. The least depth found during the low water season of 1882 was $8\frac{1}{2}$ feet, and the channel was wide and direct. That season, 5 feet depth was reported at Sulphur Spring, Forest Home, Kinney Point and Jacket Pattern, in the lower river; and $5\frac{1}{2}$ feet at three other points. The lowest

8

stage by the gauge was 1.9 feet above low water; but the lowest channel depth was found in October, when the river was at a stage, more than 3 feet above low water. The least depth found throughout the 21½ miles of river between St. Louis and Kimmswick, over which the work extended, was 8 feet. The report for 1884 says that the hurdles constructed several years ago at the up stream end of the reach, on the Illinois shore, seem to be exerting an influence over the entire area as far as Carroll Island. Though causing heavy deposits, these did not reach the desired height sufficient to shut out all the water even at low stages; but at the up stream portion of the reach, the process of building up the new bank continued satisfactorily. The height formerly reached was increased, and also the area upon which willows were growing. Upon the Missouri side the new bank had reached a height about 25 feet above low water, and was covered with a vigorous growth of new willows. The good results of the work upon the channel previously secured, were not fully maintained throughout the year. A depth of but seven feet was found upon the lower reef for a few days in September, 1883. The shoal spot soon cut out, however. The lowest gauge reading that fall was 3 feet above low water, and the least channel depth was 4½ feet at Devil's Island, 5 feet at Sulphur Springs and Jacket Pattern, and 5½ at three other points; the above quoted 7 feet being the least depth on the 21½ miles below St. Louis.

The least depth found during the low water season of 1884, on this reach, was 9 feet.

The construction account shows the entire amount spent on this stretch of five miles; the sums spent since 1879 being separately accounted for under the head of hurdles and protection.

[6] The chute between the Illinois shore and Arsenal Island, opposite Saint Louis, was closed by a dam built in 1878, and repaired in 1879. This dam had settled in 1881 three feet, but was in good condition. It fully accomplished the first result expected of it, which was to stop the erosion of the Illinois shore. The second and more important result, the deepening of the channel west of Arsenal Island, was gradually secured. Not less than 9 feet was reported for this channel in 1881. The protection of this island amounted by 1883 to 7,843 lineal feet, and since then the entire work seems to be reasonably secure and effective.

Just below Horsetail, Carroll's Island shore was protected from erosion, and these works included in the stretch already described.

[7] The locality known as Twin Hollows extends from the foot of Carroll Island to the head of Beard's Island, 3½ miles. Just above Beard's Island the width between banks was 3,650 feet. Obstructions were not always found in the same place, nor to the same degree. Sometimes the bar was opposite Twin Hollows; sometimes opposite the head of Beard's Island. At either, or both of these places, the

· 6. 1881, II., p. 1520.   7. 1882, II.. p. 1596.

channel depth was liable to be as little as four feet. In the autumn of 1879 it was 4½ feet; in 1880, no obstruction; in 1881, 6½ feet. The plan of contraction to 2,500 feet width, was begun in 1881. The line of the new bank made a small angle with the Missouri shore, at Twin Hollows; and 8,800 feet of primary, or longitudinal hurdle were built that year, connected with the shore by four secondary, or cross hurdles. Protection of the east bank of the river, to prevent any erosion from the action of the contraction works, was carried along simultaneously; 5,925 lineal feet were covered by 6,618 feet of mattress, overlapping in part. Beard's Island was also protected; 3,550 lineal feet, in all, including overlap; all mattresses 120 feet wide; and a small chute was closed. The following year the protection of the east bank of Twin Hollows was extended to a total of 8,625 feet, extending from 10 feet to 16 feet above low water. The construction of a hurdle to connect Pulltight with the head of Beard's Island was begun this year; and 2,860 feet of primary built, and two secondary hurdles begun. Beard's Island revetment was extended to 7,300 lineal feet. Subsequently, work on this stretch was confined to repairs, and the completion of that already begun. The results to the channel have been summarized above. Large deposits have been secured, and it is possible the works may not have to be carried as far as originally projected.

The next locality below this is known as Jim Smith's, which name covers the eighth and ninth shoals below Saint Louis bridge. The reach extends to Kimmswick, 3½ miles. The width varied from 4,000 to 7,000 feet; and the least depth in the channel was in 1879, 6 feet; in 1880, 7 feet; and in 1881, 6½ feet. Work began in 1882, and was extended, during the fiscal year, to a total of 8,650 feet of ·primary hurdles; and 7,600 feet of secondary hurdles, which will be six in number. The works were much injured by the ice in the spring of 1883; fully thirty-two per cent. being carried away, more or less. The least depth found during the year was 8½ feet.

The revetment of Chesley Island, on the Missouri shore, opposite these works, was a noteworthy performance. A single continuous brush mattress, 120 feet wide, and 4,305 feet long, was built and put in place. This, covering an area of 516,600 square feet, was the largest continuous mattress made, up to that time, and, had it been necessary, it could have been made longer. At the head of the island a revetment 550 feet long, and 40 wide, was made; and a hurdle 900 feet long closed the west chute.

The extensive damage, caused by the ice in 1883, was not repaired until the fall of 1884; when the hurdles were reconstructed, and the distances between the secondary hurdles diminished by the building of new, intermediate ones. The least depth found in the season of 1884 was 7½ feet, opposite Chesley Island.

The protection of the west bank of Foster's Island was begun in the fall of 1882. At the close of the fiscal year 1884, there had been built, and placed, protection 120 feet wide, and 5,284 feet long.

During this period of construction the method has remained essentially the same. A well braced primary hurdle; secondary hurdles with intervals at first as great as 2,000 feet, and afterwards lessened; well braced, and with foundation mat of brush. In the protection of banks, brush mattresses to low water line, on a graded slope; 35 to 40 feet wide, and as great as 120 feet wide; the brush to extend to a line where willows will flourish; and with a layer of stone. Above that, willows are grown.

[8] In 1881, the Mississippi River Commission, then just organized, made a formal examination of the works in this district, and Major Ernst's plans, in October, 1881; and in their report of November 25th, 1881, they fully endorsed the plans, and methods of carrying them out.

The act of 1882, required that all money spent on the Mississippi River " below Des Moines Rapids, should be expended in accordance with the plan, specifications and estimates of the Mississippi River Commission, or according to such plans of the Engineer Department, which, having been approved by the Secretary of War, may be adopted by the Mississippi River Commission." Accordingly, plans were submitted by Major Ernst, and approved by the Commission September 6th, 1882.

In the meantime, the works were progressing, and they have since continued without change in their administration.

[9] In April, 1877, was available $5,000 for the protection of the bank in Kaskaskia bend. In 1878, and 1879, this amount was supplemented by $55,000 more, and although the first work done, 1,000 feet, was carried away, 4,425 feet of revetment, in all, were placed, by the fall of 1879. The following year, 1,015 feet more of brush mattress, besides a curtain dike, 1,140 feet long.

The spring of 1881, was very disastrous to this work, the dike being carried away, the protection breached, and a channel opened into the Kaskaskia, by the combined effects of ice and flood. The work was originally intended for the protection of land, and not for the interests of navigation, which does not suffer from the failure of the work.

[10] In 1881 a special construction of a hurdle dike was made above Cape Girardeau, intending to reunite the then divided channel of the river, and to afterwards, by a second dike, direct the current so as to remove a bad bar, then in front of the steamboat landing at Cape Girardeau. Work began in August, 1881, and 2,256 feet primary hurdle, and 2,156 feet of foot mattresses were built; and 5,329 feet of secondary hurdles. The results were eminently successful, and the bar was removed in a short time, the second dike proposed, not being necessary.

---

8. 1882, II, p. 1605.   9. 1881, II, p. 1559.   10. 1882, II, p. 1649,

At Devil's Island, the chutes between it and a second island, and from the second to the Illinois shore, were closed by dams, and a dike was built from the lower end of Devil's Island during 1874-5-6. In the protection of the shore near the dike, a single mattress of brush, 495 feet long, and 50 feet wide, built in 1875, was the forerunner of the large mattresses of later years.

Above Cairo a series of spur dikes was built some thirty years ago. Deep excavations were caused by the river, between these spurs, and two or three were cut off from the bank, and disappeared. Those that remained were the subject of repair, under special appropriations in 1876, and later; and large quantities of stone were placed upon the bank between them, where most required. In the act of 1884, $50,000 additional was given for the same purpose, and a continuous mattress, in the fall of 1884, 4,563 feet long, 120 feet wide, was placed over the old work.

In 1883, Major Ernst, in furnishing an estimate, with its basis, for future work in this district, had occasion to state the following: [11] Under the new method of construction adopted in 1879, no sufficient data existed for making an accurate estimate of the cost of future work upon the river between Saint Louis and Cairo. Even now these data hardly exist, but to avoid misleading Congress, he thought best to make a new estimate, based upon such information as was on hand.

The contingencies of this work are so great that any estimate based upon the number of lineal feet of hurdle, or other construction to be built, and the cost per lineal foot, may be erroneous. The original cost per lineal foot may vary between wide limits, depending upon the weather and the stage of the river. By taking an average of several seasons an approximation to this cost may be reached, but the number of times that the silting devices may have to be repaired, or even entirely rebuilt, at any particular spot, is uncertain. Evidently the only way to reach an estimate is to take the cost per mile of some portion of the river which has been under improvement for a number of years, where the circumstances are in general the same as those to be met with hereafter, and where the works have been entirely completed. There is at present no part of the river that entirely fulfils all of these conditions.

At Horsetail, the present system began in 1879. All other works of the same general character are of more recent date; the next stretch, Twin Hollows, having been begun in the autumn of 1881. The works at Horsetail are further advanced towards completion than any others, and are the best available for basing an estimate upon. During the first two years of the work the forms of construction were largely experimental, and were undergoing modification, and their first cost was larger than it would be again under the same circumstances of wind and weather. The desired effect upon naviga-

---

11. 1883, II., p. 1186.

tion has been secured here, a wide, and deep, and direct channel having been procured; but it remained to secure these results, by further building up of the new banks, and consolidating, and protecting the new land. The cost of this item is uncertain, but Major Ernst, by giving it a value, which has since been shown to be too small, deduced an estimate per mile for the improved river as $85,000.

The reports for this district are very full and minute, giving all items of engineering methods, and incidents, as well as all financial details. Each annual report for several years has contained in it, more than the present volume has; and the student may elaborate any subject, at any place, at his will. Of late years, however, the statistics of navigation for Saint Louis have not been published.

CONSTRUCTION ACCOUNT.

	Total Cost to June 30th, 1884.
Piasa Island, dam..$	35,083
Alton, dam......	33,624
Alton, dike..	67,325
Sawyer Bend, protection..	96,804
Venici, dikes..	36,342
Arsenal Island, protection..	30,733
Closing Cahokia Chute...	116,089
Channel opposite Saint Louis.......................................	58,455
Horsetail Bar, prior to 1879, 5 dikes...........................	225,067
" " training wall.....................	80,627
" since 1879, primary and secondary hurdles..	428,982
" " protection works...................	30,699
Twin Hollows, west bank hurdles...............................	226,381
" east bank, protection............................	100,649
Beard's Island, hurdles,..	7,166
" protection	84,259
Jim Smiths, hurdles..	139,388
Pulltight, hurdles...	86,105
Chesley Island, hurdle and protection.......................	63,502
Fosters Island, protection..	44,296
Fort Chartres, dam..	36,813
Turkey Island...	24,464
Kaskaskia, protection..	66,465
Liberty Island, dam and protection.........	50,183
Devils Island, dike and dams......................................	132,398
Minton Point, hurdles..	33,436
Cape Girardeau, hurdles...	31,930
Cairo, protection...	119,869
	$2,487,134

CHAPTER XII.

RIVERS UPON THE PACIFIC COAST.

THE COLUMBIA AND WILLAMETTE RIVERS.

[1] The Columbia River rises in British Columbia, on the western slope of the Rocky Mountains, in about 50° north latitude, and 39° longitude west from Washington. Its initial course is northwesterly for 150 miles, then southerly through eastern Washington for 300 miles, receiving the Lewis and Clarke Fork, and the Spokane River, the Okinagan, Yakinna and the Snake. From the mouth of the Snake westerly, for 250 miles, to its entrance into the Pacific Ocean, it receives from Oregon the Umatilla, John Day, Deschuttes and Willamette Rivers, besides numerous smaller tributaries. Its entire length is estimated at 1,400 miles, of which 733 are navigable. Above the mouth of the Willamette it is generally rocky and rapid, and at three points, the Cascades, the Dalles, and Priests Rapids, navigation is entirely arrested. These barriers divide the river into three reaches, the upper, including the Snake River, on which navigation is continuous to Lewiston.

Between the mouth of Snake River and Priests Rapids, boats run only occasionally. The canal and locks now building at the Cascades, will, when completed, connect the navigation of the lower and middle Columbia. These two sections are unobstructed.

Upon the upper Columbia and Snake Rivers the obstructions are rocky bars, with narrow crooked channels winding through beds of rock in position, with occasional bowlders and ledges reducing the generally ample depth to two or three feet.

Before any improvements were made nearly all of them were extremely dangerous, and some were quite impassable at low water. The survey of 1868, stated that at John Day Rapids the current at low water was ten miles an hour. At Umatilla Rapids a fall of 18 feet in two miles was reported. The current at the upper entrance was stated in 1876 as 12 miles an hour. The difficult character of the rock, and the unfavorable locations have caused the work of removal to be expensive; and the cost of removal, whether by contract or hired labor, has varied from $50 per cubic yard in 1883, to $13 in 1879, and $16.50 in 1882.

---

1. 1880, III, p. 2293.

The principal obstructions above the Dalles are:

Name of Rapids.	Miles above Celilo.	When Improved.
Five Mile	5	
John Day	15-18	1873
Indian	21	
Squally Hook	24	1875-6-7
Rock Creek	28	
Owyhee	38	1877
Canoe Encampment	60	
Devils Bend	80	1874
Umatilla	85-88	1873-4-5-6-7-8-'80-2
Mill Rock	91	
Homely	110	1876-'80-1

And upon the Snake River, above its mouth, the principal rapids and other details upon the Snake are:

NAMES OF RAPIDS, ETC.	Miles above mouth Snake River.	Elevation above Sea level. FEET.	River slope per mile. FEET.	IMPROVED.
Five Mile	5.	342.	2.65	1879, 1882, 1884
"	12.69	360.	2.67	
Fish Hook	16.	368.8	2.67	1879
"	26.93	398.	3.35	
"	29.32	406.	3.03	
Long Crossing	32.	414.	3.03	
Pine Tree	35.	422.	3.03	1877, 1879
"	41.22	442.	2.28	
Monumental	44.	448.	2.28	1880, 1881
False Palouse	49.	460.	2.28	1882
"	50.	462.	6.67	
"	50.46	465.	1.16	
"	52.61	467½	2.81	
"	59.73	487½	7.22	
Palouse	61.53	500½	12.50	1882
"	61.81	504.	5.42	
"	65.41	523½	1.68	
"	69.57	530½	2.74	
Texas	71.93	537.	2.74	1881, 1882
Diamond Point	79.53	557.	2.51	
Below Penewawa	87.38	580.	2.54	
Above Penewawa	90.53	585.	2.51	
"	92.21	589½	2.94	
Almota	102.15	618½	2.94	
"	103.10	621½	4.16	
Upper Log Cabin	107.07	638.	2.81	
Little Pine Tree	124.15	686.	2.81	
Steptoe	126.21	692.	2.81	
Lewiston	137.	722.	2.81	

From Celilo to Lewiston, on the Snake, is 226 miles. Of late years steamers have gone above Lewiston, and have also gone on the Columbia, from the mouth of the Snake to Priest's Rapids, 73 miles. These rapids are a complete barrier, and cut off 150 miles of navigable river above them. The reaches of the river, above this, have 400 miles of navigable river, which will be used in time. All work done upon the river is permanent, and the removal of rocks has widened and deepened the channel, and been of great value to navigation.

The total amount appropriated for this work, since 1872, has been $216,000.

In 1884, six steamboats, of tonnage from 100 to 350 tons, navigated the upper Columbia and Snake Rivers. The principal receipts at Portland, Oregon, from this region, in the fiscal year 1883-4, were:

ARTICLES.	AMOUNT.	VALUE.
Wheat, centals..........................	2,057,746	$3,174,830
Flour, barrels.............................	175,477	769,302
Bran and Millstuff, pounds..........................	1,842,566	22,825
Wool, pounds.............................	7,539,801	1,116,665
Hides, pounds.............................	939,309	84,560
Flax Seed, sacks........................	33,222	96,075
Cattle, Sheep, Horses, and Hogs....................	26,617	383,525
All other Articles...........................	....	22,083
Total Value...............................	....	$5,669,865

During the same year the imports and exports, by steamboat, to and from Snake River points, amounted to 30,260 tons, wheat, flour and feed, and flax seed, being the principal items.

THE MIDDLE COLUMBIA.

[2]The Dalles separate the Columbia into two reaches. The total fall from Celilo to the City of the Dalles, a distance of 13.6 miles, is 56.6 feet, at extreme high water, 61.7 feet at mean high water, and 81.4 feet at extreme low water. Surveys and estimates have been submitted for a canal around the Dalles, as follows:

Low water project; canal and locks.............$7,673,495.51
High water project; Celilo Falls canal, and
    open river below that...................... 2,842,848.20

From Celilo to the Cascades is a distance of 50 miles, and the completion of the canal at the latter places will not benefit commerce until the Dalles are passable. At present freight is transferred, of necessity, at either end of this combined barrier. As yet, nothing

2. 1882, III, p. 2692.

has been done towards the improvement of the Dalles. From a study of the history of similar obstructions in the Eastern States, it must be evident that the United States will be expected to do this work, sooner or later, when expediency authorizes it, and that the Cascades canal can not be of service until the Dalles are passable. Steamers have gone down the Dalles Rapids at high water, but none have attempted to ascend.

Steamers have been in use on the Middle Columbia connecting the portages at the two rapids, and four barges were so used in 1882.

CANAL AT THE CASCADES.

The barrier known as the Cascades is the termination of the navigation on the lower Columbia. The first government survey for a navigation improvement here was made during low water of 1874. This survey developed a fall in the main rapid of 21 feet, and an additional fall of 16.3 feet distributed over the next five miles. Engorgement of the channel, high velocity, and rocky bed and plateau adjoining, were the main features. A high water examination, made in 1876, showed an upper fall of 15 feet, an additional fall of 29.2 feet, and a current of 15 miles an hour at the gorge next below the middle landing. These surveys showed that a high water canal would be a much more difficult and expensive undertaking than a low water one. A canal plan prepared from the last survey under the direction of Maj. J. M. Wilson, in charge, provided for a lockage around the main rapid at all stages, and a crib work break-water 4,050 feet in length, and 25 feet above low water was to extend below the locks. With modifications this plan was adopted by the Board of Engineers for the Pacific coast, and a contract was let October 19, 1878, for the first work of excavation and masonry. This contract remained in force until November 20, 1879. Since that time work has been carried on by hired labor under the government.

The situation is one that has made work costly, and the surroundings and circumstances are different from those of eastern rivers. Day labor has always been, and is expensive. Excessive rainfall; for the fiscal year 1880 amounted to 96.4 inches; 19.6 inches in January, 1880; and the annual rises are marked features. The rise of 1880 reached 41.9 feet at the head, and 53 feet at the foot of the canal, above low water. Consequently in excavation, and in the structure for the extension of the lower end of the canal, much trouble occurred. In this rise 1,322 feet of the protecting embankment had been built, and it was noted that the velocity of the current of the river at 12 feet from the water line was as great as 11.4 miles per hour, and about 25 feet from the water line 16.7 miles.

A Board of Engineers, consisting of Lieutenant-Colonels Stewart, Williamson and Mendell, and Majors Weitzel, Houston and Gillespie, the last now in charge, considered the subject again, in the light of the experience of construction, and reported November 13th, 1880,

[3] that the medium and high water navigation of the river just below the canal was impracticable, and that the removal of rock reefs in that section should be done before the canal work, on account of the relation therewith, and to the further benefit of low water navigation. The board also acted upon other local points, and called especial attention to the fact that final plans and estimates could not be formed until further experience gave good data. A single lock was proposed, 462 x90 feet, with a lift of 24 feet, the gate openings to be 70 feet. A guard gate at the head of the canal to exclude high water, would, with the one lock, provide for navigation to a 20 foot stage. The modified project was approved by the Chief of Engineers and the Secretary of War. The estimate made for this work in 1882 was $1,920,397. The work below the canal was continued, and the removal of rocks and reefs carried on in successive years to the improvement of navigation at stages above low water, to a reading of about 30 feet. This work has been difficult, and dangerous, and tedious, but not being finished as yet, the points depending upon it, notably the low water surface below the canal, are yet undetermined.

[4] The dimensions of the canal are as follows:

Total length, 7,200 feet; width at bottom, 90 feet; length of rock excavation, 3,150 feet; width of guard lock at head, 70 feet; depth on mitre sills, 8 feet: length of breakwater, 4,050 feet; width of crib of breakwater at bottom, 30 feet; at top, 12 feet.

In construction, the first excavation was for immediate use in building a protection embankment on the river side, between the guard gate and lock. It was to be 60 feet high; exterior slope, 1 : 2.5; inteterior, 1 : 1.5; the crown was to be 130 feet from and parallel with canal axis. The first year's flood proved it to be secure. The flood of 1880 showed that the crest should be raised, which was done. In 1881, the lock pit was enclosed by constructions, and the embankment was prolonged by a dry stone wall, with concrete hearting, to avoid contraction of water way, and yet keep flood water out of the canal prism. A bulkhead connected this wall with the wall around the lock pit. In this way floods were kept out of the canal prism and lock pit; and excavation and construction have since then gone on as fast as circumstances and funds would permit. Exhaustion of funds suspended work Feb., 1882; until after the passage of the act of Aug. 5th, 1882. The protection wall, which was to extend the canal in its lower portion, was placed in part, in excavated trench, timber cribs to 1 foot below low water, dry stone wall above it. Of this, 781 feet on river side finished in 1883. Also, 300 feet of side wall of canal prism on right side at head of canal; and excavation for opposite wall in same year, finished in 1884, to 492 feet. Left bank at lower entrance sloped and paved. Preparation of stone for lock, and excavation are in progress. Bed rock is found 5.9 feet below extreme low water at the foot of the lock.

3. 1881, III.; p. 2572.   4. 1879, II., p. 1849.

Major Gillespie was succeeded in charge by Captain Powell, in July, 1881.

Total amount appropriated June 14th, 1876, to July 5th, 1884, $955,000.

Expended to June 30, 1884:

In surveys, land, plant, and construction..........$661,495.24
River improvement below canal, drill scow, etc..   80,770.84
Tow boat in use for whole work.....................   38,985.49
Construction and outfit of bowlder dredge........   18,000.53
                                                      _____

                                                      $799,252.10

LOWER COLUMBIA.

Appropriations have been made, beginning with 1866, and aggregating $605,365 for the lower Columbia, and lower Willamette Rivers. These have been expended, in the lower Willamette, to the city of Portland, and on the bars near the mouth of that river as it flows into the Columbia. Although this is a tidal river, improvement of it has been complicated with the floods from the Columbia, principally, which carry sediment across the mouth of the Willamette to the damage of navigation over that bar. The snow flood of 1876, the highest, was 28.2 feet above low water at Portland. Continuous dredging, for years, was intended to secure from 17 to 19 feet of water over Swan Island, Post Office, Saint Helen's, and the bar at the mouth of the Columbia. This temporary improvement was supplemented, in 1873, by a dam built across Percies Slough, an arm of the Columbia, to keep the floods of the Columbia from crossing the Willamette. In 1879, pile dams with waling timbers, and filled with fascines and stone, were built across the Willamette, and Coon Island Sloughs in the same vicinity, for contraction purposes, and the head of Coon Island was revetted.

WHEAT FLEETS REPORTED AT PORTLAND, OREGON.

	1876	1877	1878	1879	1880	1881	1882	1883	1884
Arrived...............	..	...	77	56	79	79	155	67	78
Cleared	59	62	74	67	82	61	164	71	88

THE UPPER WILLAMETTE.

[5]The Willamette River rises in the Cascade Range. It is navi-

_____

5. 1880, III, p. 2280.

gable during winter and spring, for boats drawing three feet, to Eugene City, 184 miles from its mouth, and for boats of lighter draft, to Corvallis, 127 miles, since improvement, during summer and fall.

The most important aid to the navigation of the Willamette, as connecting the upper river with the lower, and with the Columbia, is the canal and locks at Oregon City, 12 miles above Portland, opened to traffic January 1st, 1873. There are four lifts, and a guard lock, with an aggregate height of 39.75 feet, in a length of 3,100 feet. The locks are 210 feet long, 40 wide, and have a depth on the mitre sill, at low water, of three feet.

The improvements made by the government have consisted of removal of drift, blasting off ledges of rock, building of wing and closing dams, and scraping off the crests of gravel bars. Besides the removal of snags, which has been carried on as high as Harrisburg, 149 miles from Portland, work on the bars has been confined to the section below Corvallis. The principal bars, distances, and work done, are as follows:

Names, and character of work done.	Miles above Portland.
Rock Island, rock excavation	16
Polally, closing dam	18
Yamhill, wing dam	40½
Union, " "	50
Beaver, " "	59½
Lone Tree " "	60
McCloskey's Chute, wing dam	62
Eola, wing dam	73
Rocky Rapid, rock excavation	76
Humphrey's Rapid, rock excavation and wing dam	84
Long Crossing, wing dam	87
Buena Vista, wing dam and scraping	90
Lower Fickels, wing and closing dams	95
Upper Fickels, wing dam	96
Pine Tree, wing dam and scraping	100
Bowers, wing dam	107
Half Moon, wing dam	110
Stewart's	111

Previous to improvement, navigation was practically suspended above the mouth of the Yamhill, 40 miles above Portland, for three months yearly, during low water; now, steamers can reach Corvallis at all times, with from 35 to 40 tons freight, and return with 100 tons.

The systematic and persistent removal of drift, and the maintenance of the dams are necessary. The Total amount appropriated 1871-1884, $145,500.

TRAFFIC THROUGH THE WILLAMETTE LOCKS.

	1879.	1882.	1883.	1884.
Freight down, tons	10,873	45,962	26,601	28,632
Freight up, tons	8,857	6,856	4,300	3,380
Lumber down, feet				800,520
Passengers down, number			4,128	4,113
Passengers up, number			3,816	4,807

Upon the Cowlitz River, emptying into the lower Columbia; the Chehalis, emptying into the Pacific; and five small rivers, with unfortunately long Indian names, flowing into Puget Sound, have been expended $44,000, since 1880, in the removal of snags. In the absence of wagon roads, these rivers furnish an aggregate of 310 miles of passable lines of communication. The lower parts of the valleys are adapted to agriculture; in the upper portions are timber, and reported veins of coal and iron. The sections tributary to the rivers, are growing in population and business.

SACRAMENTO RIVER, CALIFORNIA.

[6] A survey was made in 1879 of the Sacramento River, California, from the mouth to Red Bluffs, the head of the navigable river. The river below Colusa, 163 miles, is between sedimentary banks with sufficient power of resistance, and as far as Sacrameto City, 57 miles, feels the influence of the tide. To this point 8 feet can be carried at low tide and low stage of the water. From this point to the mouth of the Feather River, 21 miles, although obstructed by shoals, the class of boats in use are able to pass. The Feather River brings the great mass of mining detritus into the Sacramento. The next reach of 85 miles to Colusa is fairly good, and with a permanent channel for boats of 3 feet draft. *For 200 miles the lower Sacramento receives no stream, directly, from the west, and is bordered throughout this distance, on the west, by a belt of land lying some feet below the high water level; thus forming a basin for the mountain drainage. This water finds its escape, as the river falls, through sloughs, which do not however carry heavy material to the river bed. A similar basin borders the river on the east for many miles above the mouth of the Feather. Above Colusa, the river shifts more or less with every flood, and upon its subsidence the navigable channel is found in many places to have changed position. There is an alluvial and shifting bottom, lying between hard and permanent banks, some mile and a half or less apart, and about high water level. This alluvial bottom is heavily timbered. Many tributaries of high fall bring large quantities of gravel to the stream. One, Stony Creek, drains an area of 600 square miles of mountainous territory, and is subject to enormous floods estimated to be as high as 80,000 cubic feet per second, which is a good flood rate for the main river. The river slope itself is high, and the river broken into pools by rapids of steep declivity. The principal details are:

---

6. 1880, III, p. 2237.

LOCALITIES.	Distance. Miles.	Elevation. Feet.	Average fall per mile.—feet.
New York Landing, Suisun Bay.........	00	00	00
Rio Vista..............................	16.8	1.01	.061
Head of Grand Island.....................	30.6	4.66	.263
Haycock Shoal...............................	47.6	7.40	.160
Sacramento City...........................	57.3	9.35	.201
Mouth of Feather River...................	78.0	16.00	.325
Knights Landing............................	107.0	20.00	.138
Winn's Landing..............................	128.0	27.00	.328
Colusa.......................................	163.0	43.20	.177
Caldens Landing............................	175.5	54.25	.872
Princeton....................................	185.7	66.90	1.24
Foot of Pike's Cut-off....................	196.5	80.22	1.23
Jacinto......................................	203.0	92.66	1.92
Bidwell's Landing...........................	224.5	127.80	1.63
Foot Sam Soule Bar........................	231.4	134.49	.97
Head Sam Soule Bar........................	232.0	139.23	7.53
Head of Gazette Chute.....................	240.6	156.66	2.02
Merritt's Wheat Landing...................	243.4	159.29	.92
Head of Captain James' Rapids.........	252.9	186.23	2.83
Foot of Tehama Rapids....................	256.7	192.65	1.66
Head of Tehama Rapids....................	257.7	200.56	8.41
Foot of Saw Mill Rapids.......	260.5	202.77	.79
Sacramento Bar.............................	265.6	220.67	3.47
Last Chance................................	271.9	236.73	2.57
Red Bluff....................................	275.7	244.54	2.05

Below Princeton bar there is 3 feet in the channel; above that in 1879, were found 20 to 30 inches on the bars to Sam Soule bar where only 14 inches were found, and 18 to 30 inches on the bars above that. The trade of the upper river is principally the towing of wheat barges, which when loaded, carry 500 to 600 tons, on 4 or 5 feet of water. Low water lasts about three months just after harvest. The worst immediate enemy of navigation is snags which have been removed when funds allowed since 1875, the date of the first appropriation of $15,000.

The drainage area of the basin is more than 20,000 square miles, and high mountains form it in the greater part. Low water discharge at Sacramento City, 6,000 cubic feet per second. The Feather and American supply about 2,000 feet per second; 200 of which come from the American.

Efforts for improvement began with the removal of snags, but the overpowering influence upon the Sacramento and San Joaquin Rivers, of the immense quantity of mining detritus brought from the mountainous mining region by small tributaries, brought that question prominently up. Lieut.-Colonel Mendells' first report on the subject,[7] January 10th, 1881, shows the areas, causes and results to be expected, and suggested brush reservoir dams to restrain this detritus from the navigable rivers.

This preliminary report of 25 pages was afterwards replaced by a full report of Lieutenant-Colonel Mendell, dated February 2d, 1882.[8] In this the description of the mining belt, and its drainage, is supplemented by full particulars of hydraulic mining industry, viewed in all lights. The natural denudation of the California valleys is compared with that of other valleys in the world, and the influences of the mining processes explained in detail. Remedial measures are then discussed in all lights, and recommendations made for systems proposed. Maps and drawings are given, and the data upon which the report is based. The report covers 94 octavo pages, and would seem to supply all information needed for the action of Congress.

Operations for the improvement of the Sacramento and Feather Rivers continued as indicated, until the passage of the act of 1882, which gave $250,000 for the improvement of these rivers. The department decided that this sum could not be used for the impounding of detritus upon these rivers, which decision was sustained by the action of Congress in the act of 1884, which in giving $40,000, provides, "that no part of this sum, or of the money now on hand to the credit of this fund, except what may be necessary for snagging or dredging operations, shall be used, except for building a first-class dredge boat, until the Secretary of War shall have been satisfied of the cessation of hydraulic mining on said rivers and their tributaries." The building and operation of a snag boat, which was also employed to scrape the bars on the upper Sacramento, has been of material advantage to navigation, which above Colusa would have been impossible without it.

There are three lines of steamboats operating upon the Sacramento, and statistics, incomplete during some years, are given.

Tons of freight moved upon Sacramento River: 1879, 200,925; 1881, 279,659; 1882, 164,600; 1883, 107,554; 1884, 240,485.

Aggregate appropriations for the Sacramento and Feather Rivers, $445,000; of which there had been expended up to June 30, 1884, $194,963.34.

SAN JOAQUIN RIVER.

The appropriations for this river have been expended chiefly in dredging in Stockton Slough, and elsewhere within tidal limits of the lower river. Besides this, snags have been removed upon the upper river, and some brush dams built.

Nothing is given as to the commerce of the upper river.

---

7. 1881, III, p. 2485.   8. 1882, III., p. 2543.

CHAPTER XIII.

The first expression of opinion on the subject of the improvement of the low water navigation of the Mississippi River below Cairo, other than Captain Suter's report of 1875, was by a board of engineer officers constituted by the order of General Humphreys, Chief of Engineers, July 8th, 1878. This board, composed of Colonels Barnard and Tower, Lieutenant-Colonel Wright, and Majors Comstock and Suter, reported January 25th, 1879, and their report is concurred in by the Chief of Engineers.

The chief sources of information on the subject were at that time Humphreys and Abbot's report on the Mississippi, Captain Suter's report of 1875, and his map, and an early reconnoissance by Topographical Engineer Officers. Surveys of limited parts of the river had been made, and a survey was in progress under the Engineer Department. From these data a brief description of the river was given. From Cairo to the Gulf of Mexico, a distance by the river of 1,100 miles, the Mississippi flows in an alluvial plain, subject to overflow at the highest stages of the river, the width of the plain being 30 to 40 miles. Numerous curved lakes, evidently once bends in the river, and separated from it by cut-offs, and subsequent silting up of their ends, are scattered along the banks. They are sometimes miles from the river, thus indicating how unstable it is and how widely it has wandered. The river is almost a continuous series of bends, whose concave portions are alternately on the right and left banks of the river. The radius of the concave part of the bend is rarely less than a mile, and a long bend is usually about two miles, sometimes rising to three miles for bends of 180°. The water next to the concave bank is invariably deep where the thread of water of the greatest velocity follows that bank at low river. The convex side of each bend usually has a sand bar running far out from the high water shore into the river.

Wooded islands of large area are frequently included between the arms of the river, and, where it is wide, sand bars or sand islands, whose areas reach hundreds of acres, may fill the spaces between the high water banks abandoned by the water. Layers of gravel are frequently found on these bars. The material of the banks, where bluffs do not reach the river, is sand, sandy clay, and sometimes a little gravel. Many of the concave bends are receding under the action of the river, a recession of 300 feet per year not being extraordinary. On the left bank, bluffs reach or approach the river in many places; on the right bank, at but one. The following data are from Humphreys' and Abbot's report:

---

1. 1879, II., p. 1008.
9

	Area. Square feet	Width. Feet.	Maximum Depth. Feet.
Ohio River to Arkan's River, high water	191,000	4,470	87
" " " " " low "	45,000	3,400	49
Arkansas River to Red River, high "	199,000	4,080	96
" " " " " low "	54,000	3,060	56

Maximum range between high and low river at Cairo is 48.4 feet; at Memphis, 37.1; at Vicksburg, 48.3; and at Natchez, 51.5 feet. High water slope between Cairo and Columbus is 0.571 foot; low water, 0.381 foot per mile. Between Natchez and Red River Landing the high water slope is 0.266 foot per mile; low water slope, 0.150 foot per mile. In 1858 the maximum river discharge at Columbus, Kentucky, was 1,403,400 cubic feet per second, and the minimum 128,670 cubic feet. In 1881, observations under the M. R. C. gave 1,603,215 for the greatest, and 156,145 as the least discharge in a year.

In 1851-2, at Carrollton, the maximum value of the ratio by weight of sediment to water carrying it, was 1-681 in June; the value of this fraction was 1-6381 in October, a minimum. In 1879-80 these fractions were 1-793, and 1-4545 in December and October respectively. At Columbus, in 1858, these fractions were 1-670 and 1-7152. That year the highest mean velocity at Columbus was 8.47 feet per second; on June 17th the highest observed velocity in any section, 11.1 feet; and on October 16th the lowest mean velocity was 1.50 feet. In 1882 the highest mean velocity was 8,137 feet, and the lowest 1.50 feet.

At Vicksburg, in 1858, the highest mean velocity was 7.04 on May 28th; the least 3.01 on October 25th. Observations in 1882 at Helena, Arkansas, Hays' Landing, Mississippi, and Red River Landing, Louisiana, gave 6,847, 6,292, and 6,778 as maximum mean velocity readings; and 2,734, 2,284, and 2,141 feet for lowest mean velocities. The high mean velocities are ample to account for the eroding action of the river upon the banks and bed. The observations of 1882 by the M. R. C. give for—

LOCALITIES.	High Water Section. square feet.	Low Water Section. square feet.
Columbus, Kentucky............................	197,020	97,235
Helena, Arkansas................................	241,131	59,075
Hays Landing, Mississippi.....................	168,791	86,211
Red River Landing..............................	251,360	98,674
Carrollton, Louisiana...........................	167,126	131,220

In a river that rises 40 to 50 feet, there is of course ample depth at the higher stages. When the river falls, the depth may be good in the bends, and bad at the crossings. The first difficulty at a shoal crossing may be found when the river is still 6 to 10 feet above low water mark, the bars filling at high and cutting at the low stages, (as already quoted). Upon these crossings the river may shift its channel several times during low water, giving uncertain and dangerous navigation, or not being sufficiently concentrated, may not have the power to cut a sufficient channel through the high water deposit. Ordinarily the shoal places are where the river, being wide, the channel makes a crossing; but shoal water may occur in the straight reach connecting two concave bends on the same bank, when this straight reach arises from the gradual filling up of a concave bend, as sometimes occurs.

From these and other known data, some conclusions are drawn in general terms, as controlling any system for the improvement of the low water navigation of this portion of the river:

1st. There is ample depth of water at low river whenever the low water width does not exceed about 3,500 feet.

2d. There is ample depth whenever the thread of the current follows a well marked concave high water bank.

3d. Low water navigation may be bad in those reaches approximately straight, connecting the deep pools under two concave banks, if the width of the low water river in this reach largely exceeds 3,500 feet.

Another element of great importance in the problem of improving the navigation of the river, is its great instability. The general movement of the entire bed is constant and within large limits. In many of the bends the concave banks are cutting out, while the opposite points are extending. While high water velocities are ample for cutting the banks in some cases eroding or caving is largely accelerated by another action.

The river water penetrates the sand strata of the river banks to considerable distance; returning when the river falls, carrying sand with it, it allows the overlying strata to fall and cave from lack of support. Caving at the rate of 200 or 300 feet a year is not unusual, and these amounts are sometimes much exceeded. This material, once in the river, contributes additions to the bars, and points, to the injury of the navigable channel. Captain Suter's report of 1875, states that there are forty-three places between Cairo and the mouth of the Red River, where low water depth of less than 10 feet may be found, and thirteen where there may be less than five feet.

If, then, where the river is now wide and shoal, it can be aided to take a less width, a greater depth may be expected to result, provided that the bed of the river, at these shoal places, is not too hard to be worn away by the river current. Closure of "chutes," or high water channels, may be necessary, if they remain open at low water.

While no case is known of an attempt to regulate a river of the size of the Mississippi, to improve its navigation, it is possible that great silting may be produced by slight works, and inexpensive means, as shown by experiments upon the Missouri by Major Suter. At points where the river is already on the point of building up a shoal, great changes may be easily produced. The proposed works should give the low water channel a curvature like that on the river which does not produce much caving, and in straight reaches a width of about 3,500 feet.

If necessary to control the river, where the forces in operation are directly opposed, great and costly works may be necessary. In the absence of detailed surveys, it is impracticable to do more than make rough estimates; and, indeed, the river is so changeable that any detailed plans might have to be entirely changed before their execution could be completed.

Protection of caving banks will be needed to a distance above any improvement, sufficient to remove any danger of the river abandoning a route assigned to its low water channel, or of damaging the works by getting in their rear. To thoroughly regulate the river, caving, even in those bends which have deep water on the crossing below them, should be stopped, that the river may not change its position for the worse. This protection can be effected by mattresses; where the water is very deep it will be expensive; but at first it can be omitted until experience shows when and how to do it.

The main question is one of cost, which can only be determined by experience upon a bad reach of the river. Such an one as the Plum Point reach, 160 miles below Cairo. This reach is about 20 miles in length, and in many places, at low water, has a width of $1\frac{1}{2}$ miles. The case is made worse by the fact that at one place a concave bend is filling up.

Should works of a slight character,— such as hurdle work, brush ropes, and light brush dikes, whose object is to make the river drop its sediment,— and thus build its controlling works,— be used successfully, yet the cost of obtaining ten feet of water at low river, upon this reach, cannot be less than $600,000; and if more substantial works are required, this may be tripled.

This narrowing and limiting of the low water river should be attempted first by these slighter constructions.

EFFECT OF A PERMANENT LEVEE SYSTEM ON THE MISSISSIPPI BELOW THE MOUTH OF THE OHIO RIVER.

[2] A report upon this subject, by the board of five engineer officers convened to report upon the low water navigation of the Mississippi, was made January 25th, 1879, and the views and conclusions of the board concurred in by General Humphreys, the Chief of Engineers.

---

2. 1879, II, p. 1015.

GENERAL CONCLUSIONS.

1st. The levees, where they have been permanently established, do, to a certain extent, afford protection and give needed facilities to commerce and navigation, and were they permanently established throughout the river, they would doubtless develop a large additional commerce, and afford the facilities just mentioned, for its transaction.

2d. Levees have no direct action except when the water is high, nevertheless a connected levee system begins to act before the stage of actual bank overflow is reached, by closing the numerous creeks or bayous of the great swamp basins, which would otherwise begin to draw water before the higher stages were reached. This is the more usual extent of its action, and the confining in one channel of this water, in ordinary as well as flood years does, in a general way, tend to deepen the bed. On the other hand, levees closely adhering to the river banks, which in all high stages would confine the water, which now escapes into the swamps, would, by an increased current action, accelerate the caving of banks in the bends, and enhance the instability of the bed. The great obstacle to the improvement of low water navigation, and to maintaining a levee system is one and the same, for both, that is, the instability of the river from the caving of its banks.

Under the act of June, 1874, a commission, consisting of Majors Warren and Abbot, and Captain Benyaurd, of the Corps of Engineers, and Messrs. Jackson E. Sickles and Paul O. Hebert, were appointed by the President to investigate and report a permanent plan for the reclamation of the alluvial basin of the Mississippi. [3] The report of this commission, dated January 18th, 1875, necessarily contained much matter pertaining also to navigation, and which constantly recurs in connection with the improvement, or plans for it of the lower river. Therefore we find in it, data, and theories upon cutoffs, diversion of tributaries, reservoirs and outlets, besides the main subject of levees. The commission express the belief that facts do not justify a statement that the effect of embanking a river is to confine the sedimentary matter to the channel in such a manner as to ultimately raise the bed, and with it the high water mark; they also disclaim the opposite theory that by thus confining within its own bed the flood volume of the river will rapidly excavate the channel and prevent any permanent increase in the high water mark. Facts do not support either theory. Full descriptions are given of levees, history, condition and needs, and a plan of reorganization proposed. In connection are given an analysis of floods later than 1858; minute details of the levees in the different states; statistics, gauge records and flood heights of many tributaries, and all that is needed to make this report of 142 pages of great historical value when the levee question comes up on its own merits, as it will some day.

---

3. 1875, I., p. 536.

The act of Congress approved June 28th, 1879, creating the Mississippi River Commission, provides that three members, one of whom shall be the President of the Commission, shall be selected from the Engineer Corps, one member shall be from the Coast and Geodetic Survey, and three from civil life, of whom two are to be civil engineers. The President appointed Lieutenant-Colonels Gillmore and Comstock, and Major Suter; Mr. Henry Mitchell, of the Coast and Geodetic Survey ; and Messrs. J. B. Eads, B. M. Harrod and Benjamin Harrison. Of these, Messrs. Eads and Harrison have left the Commission, and Messrs. S. W. Ferguson and R. S. Taylor been appointed members. The duties of the Commission were to execute surveys of the river, and to mature such plans and estimates as will permanently locate and deepen the channel of the Mississippi River, and protect its banks; improve and give safety and ease to the navigation thereof; prevent destructive floods, and promote and facilitate commerce and the postal service. The Commission was also to make a report upon the jetty, levee, and outlet systems.

The first report was made March 6th, 1880. [4] In it the Commission says: The Mississippi River, aside from its great length and wonderful and impressive magnitude, and the energy with which it maintains its tortuous and ever changing route to the sea, ranging over a broad alluvial region of its own creation, does not appear to be characterized by any phenomena peculiar to itself.

Its waters, within the limits of the alluvial district, constantly carry large, though very variable, quantities of sedimentary matter. This is borne along by the current in such a manner that a large part of it is held in more or less constant suspension in the water, while a portion is rolled or swept along upon the bottom. An exact relation between the quantity of silt transported or moved along by a stream, and the longitudinal velocity of its current, has not been discerned. Longitudinal velocity, however, is always accompanied by motion in another direction. Without upward motion of the water there can be no continued suspension of sedimentary matter in it. When, on any given reach of the same stream, the velocity is increased, the vertical motion, or silt sustaining power, is also increased. It does not follow from this, however, that in different reaches of the same stream, or even in the same reach through varying stages of water, or different seasons of the year, the same velocity of current invariably sustains and transports the same amount or proportion of sedimentary water. No fixed relation has been discovered between a volume of water and the amount of sediment in it, for any observed velocity. The supply of earthy matter is very irregular, varying greatly, from various causes. Any reduction of velocity, by lessening the sustaining power of the water, will tend to cause a deposit of solid earthy matter. Conversely, if the velocity be increased from any cause, a greater amount of silt will be thrown into suspension.

---

4. 1881, III., p. 2719, et seq.

These general principles may be briefly stated as follows:

If the normal volume of water in a silt bearing stream flowing in an alluvial bed of its own formation be permanently increased, there will result an increase of velocity, and consequently of erosion and silt bearing power, an increase in area of average cross section, and an ultimate lowering of the surface; and, conversely, if the normal flow be decreased in volume, there will ensue a decrease of velocity of silt transporting power, and mean sectional area, and an ultimate raising of the surface slope.

Whatever seems needful to be said on the "outlet" system, which, being one of diffusion and waste, and not of concentration, does not commend itself to the Commission, will be found embodied in this report.

Levees have never been erected upon the banks of the Mississippi except for the special purpose of protecting the alluvial lands from overflow. There is no doubt that the levees exert a direct action in deepening the channel and enlarging the bed of the river during flood periods, by preventing the dispersion of flood waters. There is reason to believe, that during the period when levees were in their most perfect condition, from 1850 to 1858, the channel of the river was better, generally, for purposes of navigation, than it has been since. While it is not claimed that levees are necessary for navigation, it is believed that the repair and maintenance of existing levees will hasten the work of channel improvement, and it is certain that they give ease and safety to navigation, and promote and facilitate commerce, by establishing banks and landing places above the reach of floods.

The bad navigation of the river is produced by the caving and erosion of its banks, and the excessive widths and the bars and shoals resulting directly therefrom. It has been observed in the Mississippi River, and is indeed true of all silt bearing streams flowing through alluvial deposits, that the more nearly the high-river width approaches to uniformity, the more nearly uniform will be the channel depth, the less will be the variations of velocity, and the less the rate of caving to be expected in concave bends. Uniform depth joined to uniform width, imply uniform velocity; and this means that there will be no violent eddies and cross currents, and no great and sudden fluctuations in the silt transporting power of the current. There will therefore be less erosion from oblique currents, and no formations of shoals and bars produced by silt taken up from one part of the channel and dropped in another.

The work to be done, therefore, is to scour out and maintain a channel through the shoals and bars, and to build up new banks and develop new shore lines. It is believed that this can be done by contracting the low water channel way to an approximately uniform width of about 3,000 feet. The channel should be maintained in its present location, and no attempt should be made to straighten the river, or shorten it by cut-offs. The methods of construction will be the hurdle, the open dike, the continuous brush mattress, curtains of wire or brush netting, and other forms of permeable brush dikes, jetties, or revetments. That these methods are practicable is shown by the works already executed upon the Mississippi and Missouri Rivers.

The plan already proposed by the Board of Engineers for the improvement of Plum Point Reach is substantially adopted for the initial works proposed there. Estimates intended to cover the construction, but not the maintenance, were submitted, amounting to $736,000 for Plum Point Reach, 38 miles long. Five other localities were also selected for initial work.

Extensive and minute surveys were inaugurated and carried on over the river, resulting in an accumulation of data as to topography, levels, hydrography, discharge, and velocity of current, sediment, and the shape and movements of the bottom of the river, which altogether compose a comprehensive knowledge of the habits and characteristics of the stream. Equally extensive and accurate information has never been acquired as to any other great river of the world. Some of the facts thus gathered will now be quoted.

A party made continuous observations over Plum Point Reach for a year, ending October 20th, 1880:

Discharge at Fulton.	Millions of cubic feet.
November 13th to 30th, 1879................................	324,258
December, 1879..	1,439,809
January, 1880...	2,421,082
February, 1880..	1,709,941
March, 1880...	2,838,360
April, 1880..	1,991,222
May, 1880...	1,625,955
June, 1880...	1,179,319
July, 1880...	1,564,509
August, 1880..	594,588
September, 1880..	557,255
October, 1880...	319,241
November 1st to 13th, 1880	197,151
Total..	16,762,691

The maximum discharge was March 24th, 1,117,425 cubic feet per second. The low water discharge, 83.260. The vertical velocity curves show that they vary with the force and direction of the wind. For a very heavy down stream wind the curve might become a straight line with considerable inclination to the vertical; for an up stream gale, the retardation of velocity near the surface might be sufficient to render the curve nearly vertical below mid depths.

The ratios of mean to mid-depth velocities were quite variable. In the observations of February 1st velocities exceeding 5 feet per second were found at three feet above the bottom. Marked irregularities in single velocity observations indicate pulsations already noted upon other rivers. Owing to these, surface velocity curves presented irregularities,

As to sediment, it is noted that no relation is seen to exist between the mean velocity of the river and the discharge of sediment. The difference between the amount of sediment discharged by the Ohio and that coming from the Mississippi will account for much of the variations in amount of sediment passing; the more especially when the rivers are in flood at different times.

The proportion of sediment in river water is as follows:

The numerator is unity; the denominator is given in the table.	By weight.	By bulk.
Mean of January observations	1,266	2,405
" February 1 to 26	1,428	2,713
" April	1,205	2,290
" May	1,234	2,334
" June	1,000	1,900
" July	581	1,054
" August	1,075	2,042
Maximum July 10	431	819
Minimum April 2	3,704	7,037

The proportion of sediment transported at mid-depth and bottom, is about the same, at all stages. As the river rises, the proportion found at the surface increases, while that at mid-depth and bottom correspondingly decrease. While the sediment is quite evenly distributed over the width of the river, the maximum amount is found in mid-stream. The total discharge for 250 days was 696,730 millions of pounds; the yearly ratio would be 1,017,226 millions of pounds, or enough to cover a square mile 308 feet in height.

Dredging observations showed that in the bed the chief ingredients were coarse sand, varying from 57 to 75 per cent.; fine sand 12 to 37 per cent.; gravel from 5 to 15 per cent.

The amount of caving banks upon both sides was determined accurately, over ten miles of reach, for the ten months ending October 1st, 1880, as following:

Volume of caved land, cubic feet	119,261,000
Area, acres	72.6
Mean depth, in different sections, feet from	21 to 47

Of the total volume of caving, forty-five per cent. was due to local causes. Fifty per cent. occurred at stages exceeding twenty-five feet above low water, when the average rate of caving was three times greater than for lower stages.

Whether the rapid shoaling of the river at the discharge section, at the high stage of March, was general, over the Fulton section, can not be ascertained, as surveys were not taken at short intervals. While they show a fill from the beginning to the top of the March rise, they do not show whether this fill occurred gradually, or whether it was more rapid at the higher stage.

A party made similar continuous observations at Lake Providence, Louisiana, over a reach of nearly ten miles in length, from October 31st, 1879, to October 5th, 1880.

A summary of some of the conclusions reached from these surveys is as follows:

1st. The bottom of the river is in an unstable condition, and constantly in a state of motion. When, from any cause, the velocity of the current is suddenly increased, the most rapid erosion takes place; and the greatest deposit occurs, when the velocity is suddenly decreased.

2d. The transverse ridges forming the bed are most marked in deep water and rapid current, attain a maximum size and rate of progression at the highest stage of the river when at a stand, and have the least dimensions and rate of travel at low water.

3d. The heavier material is moved down stream and over the crests of these waves, and at high water the amount is greatest; or when the velocity has been suddenly accelerated.

4th. Changes in the form of sand waves are gradual, except when rapid changes in velocity occur.

5th. A large amount of material is transported along the bottom. The amount as measured by the progression of the waves is believed to show only a small part of that actually in motion. Soundings over 23 different cross sections showed that an increase of area occurred, as the river passed from low to medium stage, caused by erosion of the bottom, at the upper 17 sections; but this was not the rule for the lower 6, owing to local causes.

When the river overflowed its banks in April over the entire reach, a marked decrease of area occurred at nearly every section, and when the river regained its banks in May a scour of the bottom took place, and the material deposited was rapidly eroded. Caving of the banks occurred at both high and low water stage. Between November and May this caving took place to an average width of 125 feet, over a length of 3,800 feet on the left bank. At another part a caving to an average of 103 feet in width occurred. On the right bank a caving over a distance of 5,000 feet to an average width of 175 feet, was the greatest amount reported for the reach. Caving also occurred at other points to a lesser degree.

From December, 1879, to October, 1880, continuous observations were made at Carrollton, Louisiana, a short distance above New Orleans, and about 100 miles from the Gulf. The river is here characterized by short sharp curves and long straight reaches. Its width averaging about 2,300 feet, is less and more uniform than at points above, and the mean depths are correspondingly greater, though by no means uniform, the variations being nearly as great as between Cairo and Memphis. The banks are stable, and are raised above the floods by levees, which run close to and are mainly parallel with the high water lines. The oscillation is about 16 feet, the slope small, though the range of its variations is considerable. The range of mean velocity does not differ greatly from that at other points, being only a trifle less than that noted at Fulton, 60 miles above Memphis, during the same period.

The average gauge and total discharges are given thus:

DATE.	Average gauge for month, in feet.	Total discharge for month, in millions of cubic feet.
December 1879	3.28	12,406
January 1880	9.86	21,211
February "	10.56	18,727
March "	12.61	23,648
April "	13.48	24,663
May "	12.87	24,309
June "	7.51	14,407
July "	7.65	15,237
August "	3.31	9,961
September "	1.91	7,625
October "	1.05	5,884
November "	1.19	6,074
Average stage for the year	7.1	

Average discharge for the year was 582,000 cubic feet per second.

The means of the surface, mid-depth, and bottom observations of sediment have the ratios 100, 144, and 183, respectively; but individual observations have a wide range. Moderate change at the bottom, either by scour or filling, is generally accompanied by increase of bottom sediment, while if the action be intense, the mid-depth is also affected. The June rise coming from the Missouri River brings an increase, and an actual discharge of sediment nearly equal the maximum. At other times the variations in the amount of sediment fairly indicate the general action going on in the bottom of the stream.

The movement of sand waves along the bottom indicates less than one per cent. of the total sediment passing in the whole river, as being carried in that way; and ninety-nine per cent. of the annual discharge of solid materials will be detected by the usual observations for sediment in suspension. From the plate showing the gauge record, velocity, and sediment curves, it is seen that from December 21st, to January 31st, an increased gauge reading, and increased velocity, were accompanied by increased sediment only at first, for ten days;

later on, a deposition, or diminished amount, is shown. A long period of nearly uniform velocity and gauge, from March 10th, to May 20th, showed a large decrease, at first, in bottom sediment, and later on, a rapid decrease; which preceded a rapid fall in the gauge, and in the velocity; this fall not being accompanied by any change in sediment. After this date the curves show close relations between the three quantities; the sediment in July increasing more rapidly for a slight increase of gauge, and velocity, than shown by the record for the year.

In the portion of the river under consideration the efforts of the stream, and the form of the bed are in comparative equilibrium. The variations, from day to day, are relatively small, and during the river cycle are compensatory. A diminution in discharging capacity of the channel takes place on the rise from low water, at a point somewhat above mean stage. The capacity thus lost is gradually recovered during the low water period. The first effect of the approaching flood is to impede its own channel, and the impediment outlasts the flood. In its low water condition, the rise and fall of the river are accompanied by a scour, and fill on the bottom, small in amount, but definite. The width of 2,500 feet at Labarre is great enough to permit a pernicious shifting of the current.

During a survey from Saint Louis Landing to Arkansas City, a distance of 110 miles of river, the following facts were ascertained in 1881-2.

. 1st. The highest land is near the river, usually within 650 feet of it.

2d. The right bank is higher than the left bank by an average of 0.6 foot.

3d. The highest land between the river and the levees, which are about a quarter of a mile from the bank, averages 0.45 foot higher than the highest land inside the levees.

From Arkansas City to Donaldsonville, 499 miles, subsequent surveys in 1882-3 confirmed these results, until a point near Baton Rouge was reached. From this point down the levees are very close to the edge of the bank, and the land outside the levee is very often three or four feet higher than that on the inside. This indicates a deposit of that amount since the levees were built. Sometimes this deposit reaches nearly to the top of the levee.

The phenomena of cut-offs, effected upon the river by erosion, have been especially described by Major Suter in his report of 1875. The cutting of the bank upon the concave shore, where the bends overlap each other, reduces in width the neck of land separating these two concave bends. In the course of time the dividing neck becomes so attenuated as to be no longer able to sustain the pressure of water against it from that side where the river is highest, and as the nature of the materials of which the bank is composed allows more or less water to leak through and wash out the sand layers, finally the whole mass crumbles, and a wide breach is formed through which the river pours with resistless force. Davis's, one of the most recent of these cut-offs, and also the largest, occurred in 1867. It cut off Palmyra Bend, 18 miles below Vicksburg, a bend which was eighteen miles long, while the distance across the neck was only 1,200 feet. The slope of the river at that time was about 0.3 foot to the mile; therefore the difference of level on the two sides of the neck was about 5½ feet. When the river broke through this neck, it was as the giving way of a tremendous dam, and the effect of the immense flood volume of the river pouring through this gap can hardly be imagined. The roaring of the waters could be heard for miles, and in the course of a few hours a channel a mile wide, certainly over a hundred, and probably nearly two hundred feet in depth, had been excavated. Even then it was many weeks before the velocity of the current had sufficiently abated to allow boats to use the new channel.

It is clearly evident that in such cases as this, the stream cannot long remain in a condition so different from its normal regimen; the length by which it has been shortened must be regained, so as to restore the usual slope, and this can only be done by the elongation of the bends above and below the cut-off, and this is found to be the case. On other streams this phenomenon usually obtains, but generally the mode of action is different. On the Missouri the movement is down stream of the bends, by the building up of bars on the down stream side, as the upper is abraded. [5] Captain Suter reported on February 6th, 1871, on the threatened cut-off at Vicksburgh. Tracing the changes that had occurred since 1828, and noting the rate of erosion, he states that the width of the peninsula, or neck, was 9,100 feet in 1828; 7,200 feet in 1858; in 1865 it was 5,760 feet, and in 1866 and 1867 the erosion was very great and erratic. From 1869 to 1870 the maximum cutting was 1,500 feet, and that at the narrowest part of the peninsula. That fall the least width was 2,500 feet, but this was 2½ miles below the former place of cutting, the site of the military canal, which had increased in width to 6,833 feet. Captain Suter recommended immediate revetment of the shore for nearly four miles at an estimated cost of $2,745,535, and stated that it was not improbable that the cut-off would be made within two or three years. No action was taken by Congress, and the cut-off took place in 1876. Terrapin Neck cut-off occurred in 1866, the width of the neck being 450 feet at the time. A cut-off, shortening the river about 11 miles, occurred in 1874, 270 miles below Cairo, near Commerce, Mississippi, and one called Centennial cut-off, 205 miles below Cairo, in 1876. From special examinations made as to the changes in the river between Grand Gulf and Donaldsonville, 250 miles, during 55 years, we learn that for more than half the length of the reach the present bed of the river is entirely outside the limits it held in 1828, and in some cases

5. 1871, p. 378.

it is several miles from it. It is specially noted that the islands have moved down stream, the bends have moved down stream, and the preponderance of the movement of the river as a whole has been to the west. The bends have grown longer. From Grand Gulf to Fort Adams, 122 miles, this lengthening amounts to about 4 miles. Below Baton Rouge the changes are slight, in comparison.

On May 11th, 1884, a cut-off occurred across King's or Cole's Point, 666 miles below Cairo, which shortened the river about 12 miles. The neck was 2,200 feet wide when the surveys were made in 1882. The effects produced upon caving bends above and the levees below, were reported as very injurious.

The maps of the Mississippi River published by the commission, shows the following as possible cut-offs in the future:

LOCALITIES.	Miles below Cairo.	Least wi'th of neck. feet.	River wo'ld be short'd. miles.
Wildwood Point	409	6,000	14
Georgetown Bend	447	3,900	8.5
Miller's Bend	461	4,600	12.5
Spanish Moss Bend	470	6,200	14

If either of the two last occurred, the other would probably not take place.

From the lines of levels run in 1880 to 1883, across the alluvial bottom of the Mississippi, and of which plates are published in the report for 1883, the following notes are made:

1st. The line across the St. Francis bottom, from Stewart's landing, 90 miles below Cairo, show the Mississippi banks exceeded in height by only one stretch of one and two-thirds miles on the east of Open Bayou, for a distance of 24 miles on the west. High water of 1882 overflowed the east or left bank for more than three miles. The Saint Francis banks, 34 miles west, are only just higher than those of the main river.

2d. From Fulton, 175½ miles below Cairo, west to Crowley's Ridge, for a distance of 43 miles, the high water of 1882 passed over every foot of ground, which fell off from the river to the Saint Francis, 32 feet in all.

3d. At Glendale, 306 miles, a line 22 miles east to the highlands, carried the high water of 1882 over the whole. This line crosses Trout Lake, Boar Lake, Arkansas Bayou, White Oak, Coon, and Coldwater Bayous. The slope of the land is from the Mississippi, with two exceptions, which alone exceed the river bank in height.

4th. A line from Grand Lake, 511 miles below Cairo, to Bayou Macon, shows a fall of more than 6 feet in 5 miles; thence crossing to the cut-off, a basin of 17 miles in width was overflowed in 1882, the greater part below the level of the river banks.

5th. From Lake Providence, 542½ miles below Cairo, the high water of 1882 overflowed the 43 miles to Yazoo City, and at no point on the line is there an elevation equal to the river bank. This line crosses Steeles Bayou, Deer Creek, Little and Big Sun Flower Rivers and Panther Creek, all in depressed courses, and lower than the banks of the Mississippi.

6th. From Saint Joseph, Louisiana, 648 miles below Cairo, a line 25½ miles to the west, show a slight rise of land as high as the levee, but the whole distance was overflowed in 1882, the bed and banks of each stream being less in height as the line leaves the river. Van Buren, Little and Big Choctaw Bayous, Tensas River and Bayou Macon are crossed.

7th. From Knox Landing, 753 miles below Cairo, the land falls about 15 feet in 7 miles to the Red River. West of that the level for several miles is about that of the Mississippi bank, but the high water of 1882 overflowed it to Old River, more than 26 miles.

These all show, as need hardly be proved, that the Mississippi River builds its own bed, and its overflow deposits rest near their place of leaving the river.

CHAPTER XIV.

LOWER MISSISSIPPI RIVER CONSTRUCTION.

The first work of construction done upon the lower Mississippi, was in consequence of the Vicksburg cut-off of April 27th, 1876. The break occurred opposite the lower portion of the city, and all of the wharf front, elevators, and warehouses, were left upon a lake, in which the remains of the old peninsula appear as an island. Delta Point, opposite Vicksburg, began to recede rapidly, the bottoms below the Vicksburg Bluffs were deeply eroded, and a new bend developed there, while the whole river traveled down stream, and receded from the city. The harbor proper of Vicksburg, formerly the finest on the river, began to silt up, and soon deteriorated so much that a board of engineer officers was convened to devise a plan of improvement, reporting January 22d, 1878. The board recommended:

1st. That the Delta Point be protected, by suitable revetment, from further recession.

2d. The building of a certain dike.

3d. The dredging out of the inner harbor, and its maintenance by dredging.

4th. If these measures proved insufficient, to divert the Yazoo River so as to flow in front of the city.

Appropriations made in 1878-9 and 1880-1, amounting to $229,000 were expended in succession in revetting Delta Point, which proved to be a difficult task, and more expensive than first estimated. By

February 10th, 1883, about 10,000 linear feet of revetment had been built in successive years, of which much was in renewal of that carried away, and since that time it has been held, by constant repair against the attack of a strong current. It is reported that this has proved to be the most difficult and expensive revetment executed by the Commission, costing about $13.37 per running foot. In 1883, a contract was let for dredging a basin for the harbor of Vicksburg, and between April 5th and September 11th, over 350,000 cubic yards of mud were removed. The work was abandoned, however, at that time, because of the instability of the dredged basin, which filled up from the sides, by pressure upon, from the side masses, and rising of the bottom, and from sedimentary deposit. No further dredging has been done. The steamboat landing has been removed to deep water below the town, when low water requires it.

The total expenditures on Vicksburg harbor to September 30th, 1884, had been $350,984.31; of which $303,230 had been expended upon Delta Point.

Work designed to protect the water front of Memphis from erosion, was inaugurated in 1878, with an appropriation of $46,000. During the three following years $67,000 more were appropriated; and with these sums 3,125 lineal feet were protected, up to June 30th, 1881; and during the ensuing year 11,460 square yards of revetment were placed. After June 30th, 1882, this work passed under the Mississippi River Commission. This work was continued and maintained with good success, until 1884, when extensive caving occurred along the line of the revetment in many places, with much injury to buildings along the front. In repairing the damage, difficulty and loss were encountered in sinking mattresses 300′, and 150′ wide. After 1882, the expenditures are included with those for Memphis Reach, which will be found elsewhere.

By the act of March, 3d, 1881, the sum of $1,000,000 was appropriated for beginning the work of construction upon the lower Mississippi, and until the passage of the act of August 2d, 1882, the commission was in direct charge, having Capt. J. B. Quinn Corps of Engineers as disbursing officer at Plum Point Reach, and Lieut. W. L. Marshall Corps of Engineers at Lake Providence Reach. At the request of the commission the system was changed by the act of 1882 so that engineer officers were placed in charge of all the work of construction within certain districts of the river; in execution of plans approved by the commission. Supplies, and plant for use in construction with hired labor under the district officers, were bought either by themselves or by the executive officer of the commission, when duly authorized by that body. The districts were as follows:

First. Cairo to Island No. 40, 220 miles, Capt. J. G. D. Knight.

Second. Island No. 40 to mouth of White River, 180 miles, Capt. A. M. Miller.

Third. Mouth of White River to Warrenton, Mississippi, 220 miles, Capt. W. L. Marshall.

Fourth. Warrenton, Mississippi, to head of Passes, 484 miles, Maj. Amos Stickney.

The act of 1882, appropriated $4,123,000 for the whole. Up to November 25th, 1881, the operations of the commission were directed principally to the carrying on, and study of surveys; to the preparation of plans; to organization, and to the purchase and construction of steamboats, grading boats, pile drivers, barges, and other floating plant; to the purchase of materials, tools and supplies, and in beginning work at Plum Point and Lake Providence.

PLUM POINT REACH.

Though this reach, as defined by the commission, is 40 miles long, work of improvement has been limited to a stretch of 13 miles, from Gold Dust Landing down to Yankee Bar. Above this stretch, 2,694 linear feet of mattress 138 feet wide, were placed in 1882 at Ashport, and so far have served to prevent caving. The principal features of construction will now be briefly described, beginning with the Gold Dust dikes and going down stream.

For the prosecution of the work in 1883, there were the following boats, known as the "plant":

Pile drivers, twenty-three, all designed to sink the piles with a hydraulic jet assisted by a heavy hammer (with 30 to 45 feet leads), when stiff soil required. From twenty to twenty-three piles a day have been driven under favorable circumstances, but otherwise twelve might be a fair daily average. While used in 45 feet water and a current; yet driving is difficult in more than 30 feet. Total cost of these pile drivers $108,805.

Mattress boats, eight; one 212 feet long, for 200-foot mattress; three 100 feet long, for 100-foot mattress; one 175 feet long, for 150-foot mattress; the others smaller. Cost, $43,310. Quarter boats, sixteen; to accommodate the laborers, and supply all possible comfort to those who are subject to the annoyances, heat, and malaria of the location. Cost, $58,320. Graders, two. Cost, $64,260. These are each fitted with three boilers, each 22 feet long, by 40 inches diameter; cylinders 24 inches by 18 inches, and 36 inches; pumps which deliver 2,000 gallons a minute. Two 1¼-inch nozzles, and one 1¾-inch nozzle, are ordinarily used; pump pressure, 110 pounds; nozzle pressure, 80 pounds. One grader has removed from 1,800 to 4,000 cubic yards of earth per day, at an average cost of less than three cents per yard. To supply material to various parties, are used fifty barges, each 100x25 feet, at a total cost of $82,045. One steamboat, $11,740; three steam tugs and launches, $14,200; and two derrick boats, $9,900; and one machine boat, $8,200, complete the list of large boats. Seventy-nine small boats were also in use, at a total cost of $10,995.50. The total cost of all plant in use was $398,739.50.

Pile dikes were driven at first in parallel lines, the tops sprung together and tied. Afterwards a single line of vertical piles was

10

driven, braced with piles driven at an angle. Later, the line of supporting piles was driven at distances from the main line, varying with the depth of the water, and the two lines were braced together. In May, 1883, three lines of piles were used, the width being twice the height, the whole well braced. In the front row two, three, or four piles were used in a cluster, when the depth was 20, 30, or 40 feet or over. In deep water double or treble piles were used in the back rows.

Foot mattresses, 50 to 100 feet wide, are used for the protection of pile dikes to prevent scour. They are made as other mattresses, and sunk by starting at the head and keeping the crease in the general direction of the current, loading with stone when necessary.

Shore mattresses, for the protection of the banks, are made of willow brush 2 to 5 inches at the butt 40 feet long; woven on poles 7 to 8 feet apart, or on a wire basis. They are made 100 to 150 feet wide and in continuous lengths, and as much as 170 running feet a day have been made. High water revetment is made of a frame of stout poles, a layer of brush, and a second frame of poles on top, well wired at the intersections; the whole well wired and sewed through and through. Tipped and grillage mattresses were used with the dikes.

Brush was furnished by contract, and the labor done by men paid directly out of the appropriation.

The soft, permeable and unstable soil, in which the dikes were placed; the oscillations of water with flood and drift, or low water, and undermining; high velocity of current and great depth of water, were the elements against which the work of construction contended.

The Gold Dust dikes were designed to continue in a gentle curve the Tennessee shore, in the direction of the large bar known as Elmot bar; and by causing a deposit behind the works to stop the flow between Elmot bar and the Tennessee shore, thus narrowing the river, and concentrating it to the one channel west of Elmot bar. The main dike was to be connected with the shore by cross-dikes, used as already explained in the river below Saint Louis. Prior to December 1st, 1882, the line of the main dike was built during that year to a length of 5,200 feet. At that date the building of a foot mat 100 feet wide in front of the main dike was begun. Up to February 21st, when work was suspended by high water, this mat had been built and sunk in two sections, one 1,713 feet long, and the other 2,951 feet.

On December 10th, cross-dikes Nos. 1 and 2 were begun, and later the dikes 3, 4 and 5. In all 4,899 linear feet of dike were built. Foot mats 738x50 were built for dike No. 1, and 1,270x50 for No. 2. Wattling 10 feet wide was placed on the main dike for its entire length.

A sudden rise, with large amounts of drift, did much damage in breaching these works, and yet caused extensive deposits in places. Work was resumed April 16th, the river having a 28 foot stage.

The western end of the main dike was extended, and cross-dikes 3, and 4, to their junction with the main dike. May 1st, a grillage mat, in sections 60′x 35′, was built, and sunk with stone, beginning at cross-dike 4. A foot mat, in front of the main dike, was begun, but drift forced the suspension of its construction. About June 1st, high water again forced a suspension of work. Cross-dike No. 5 was then begun. From April 16th, to June 1st, there were built:

Main dike..........5,200 linear feet; also 2,200 linear feet mat.
Cross-dike No. 3, 1,900   "   "   850   "
    "   No. 4, 2,800   "   " 1,600   "

Wattling six feet high was placed on 2,600 linear feet of these dikes.

This rise caused much damage, by scour and drifts, but large deposits formed in rear of the main dike. When the water fell, operations were resumed in closing gaps, strengthening the pile dikes, putting in tip mats in front of cross-dikes, where scour was to be stopped, and bank protection where the dikes touched the shore.

The completed work was then reported to be as follows:

DIKES.	Dike. linear feet.	Grillage or Foot Mat. linear feet.	Wattling. linear feet.	Tip Mat. linear feet.	Bank Protection & mattress. linear feet.
Main Dike............	9,457	8,311	1,600	1,325	410
Cross-Dike No. 1....	700	200	....	....	....
" " 1 a.	630	225	....	....	....
" " 1 b.	200	....	....	....	....
" " 2....	700	350	....	....	....
" " 3....	2,910	2,810	850	....	303
" " 4....	4,400	2,630	1,020	300	340
" " 5....	3,400	3,100	450	1,505	....

The total number of linear feet of pile dike driven to form these dikes is reported, at this time, to have been 35,083 feet; at a total cost of $406,624.81; and an average of $11.60 per foot. These works had then cost about forty per cent. of the total expenditures upon Plum Point reach.

Openings had been left in the upper lines, to allow access to the lower unfinished ones, and these caused scour, and loss from drift, in 1883-4. The pile work of the main dike was finished in 1883. The foot mat work was much interrupted by high water, but finally completed in the fall of 1884. Heavy fills occurred behind the main and cross-dikes, at some places rising to thirty feet above low water level. At the Tennessee end of the main dike the material was deposited as high as the main shore, and a level bar extended out some distance behind the main dike.

Completing the cross dikes was slow work, owing to the difficulty of getting material when wanted, and boats could not always be towed by steamer where wanted, but before high water they were finished, except in the openings left as a passage way. The laying of grillage and protection mats and the repairs of damages have been principal items in 1884. The officer in charge thought the work, as projected, would be completed before high water in the winter of 1884-5, and work, beyond repairs, has not been done since then. Existing gaps were closed before January 1, 1885, and continued filling occurred, and a total deposit behind the works of 3,000,000 cubic yards is reported. To October 1st, 1884, the total amount expended upon the Gold Dust dikes has been $594,892.06, covering 38,974 linear feet at an average cost of $15.26 per foot. These works have cost 31 per cent. of those of the whole reach, up to that time.

FLETCHER'S FIELD REVETMENT.

Directly opposite the Gold Dust dikes a revetment to protect three miles of bank was projected as a necessary consequence of the first named works, and construction began May 26th, 1884, and continued until January, 1885, when 6,700 linear feet were protected. During the preceding year caving to a maximum width of 700 feet had taken place in this bend. In construction, difficulty was encountered in the removal of an interlaced mass of fallen timber, which was done by the aid of one of the United States snag boats. October 1st, 1884, the revetment had cost $88,584.80 for 6,150 linear feet, or an average of $14.40 per foot. Caving is taking place where there is no protection.

OSCEOLA BARS.

By connecting the Arkansas shore with the upper Osceola bar, by a dike making a gentle curve, and by continuing the line for a new bank down from the upper to the lower Osceola bars, and finally in time from the last to Bullerton Towhead, a new western line for the bank of the contracted river could be secured some seven miles in length. This new line would for long distances be more than half a mile distant from the old Arkansas shore, and would have a less, but still great distance, for the greatest part of its course. Between Osceola Bar and Bullerton Towhead ran the main channel which was to be closed, and forced to pass to the east of Bullerton Towhead.

Work began April 8, 1881, for the closure of the upper or Osceola chute. When stopped by high water, 4,500 linear feet of dike and 570 feet of foot mattress had been built. Drift and undermining at low water carried all but 300 feet of the dike away. That fall the dike was rebuilt, and a foot mat, 3,715 feet long, laid to complete that for the entire dike. Though structurally weak, and damaged from time to time so that by the end of 1884 only about one-third remained, yet the deposits behind this dike have been large.

A secondary dike was built, in 1883, from the Arkansas shore to the head of the upper bar, a length of 1,355 feet. Excessive drift prevented the proper mattressing of the foot of this dike. An area of twenty acres of closely wedged drift was accumulated in front of this dike in a few months in 1883. Although producing scour, the drift also checked the current. The dike held, and large deposits occurred above and below.

At its lowest point the chute was dry in 1884, at a 10-foot stage, while the fill is nearly as high as the tow-head, for two-thirds of the distance across, and is covered with a thick growth of young willows.

Up to October 1st, 1884, the bank protection of the upper bar, for a length of 3,100 feet, cost $27,081.64; the upper dike cost $48,976.31, for 7,315 feet; and the cross-dikes $169,576.68, for 9,437 feet.

The protection has suffered no damage.

Between the upper and lower bars a dike, called the middle dike, was begun in 1882, and pushed to a length of 3,309 feet by December. The 100-foot wide foot mat did not prove sufficient protection in this case, and by November 1st, 1883, only 1,059 linear feet remained; which the floods of 1884 still further reduced to about 600 feet in length, by October, 1884. The middle dike cost $49,-300.93, for 3,773 feet.

Secondary, or cross-dikes, from the Arkansas shore to the two bars, and from the upper to a secondary bar, were built in 1883,—to check the current of these chutes,—to a total length of 9,437 feet; met the usual chain of accidents, but, with repairs, have been held, and are doing their work in securing deposits.

Considering the dikes as a whole,—being intended to close Osceola Chute,—it is stated that they may be called a success.

The revetment of the lower Osceola bar, as necessary to hold this line, was begun in January, 1883, and continued, at intervals, with varying success, until January, 1885, when the total revetment covered a length of 4,500 feet; and 750 feet more was covered with mattress and partial revetment, and 150 feet remained unfinished.

To October 1st, 1884, the lower bar revetment had cost $88,584.85 for a length of 4,300 feet.

The entire Osceola system had cost $383,340.31, or about twenty per cent. of the expenditures for the whole reach.

OSCEOLA-BULLERTON MAIN DIKE.

This dike was intended to connect these two bars, and by crossing and closing the main channel of the river, deflect the current to the east of Bullerton towhead, and force it to excavate a new channel along that line. Work began at Osceola lower bar September 1st, 1882, and piling was carried to a depth of 25 feet of water, and through a distance of 2,117 feet. The foot mat, 70 to 96 feet wide, was 2,197 feet long; and wattling was carried to the 10-foot level.

Deposits were immediately formed behind the dike, and the new channel began scouring where wished within two months. The dike was also begun at the Bullerton towhead end and pushed to a length of 800 feet during the same interval.

By December 1st, 1882, but 1,200 feet of the first or north dike was standing. Construction was resumed March 12th, 1883, and only by April 13th, the dike was completed, all but two openings, one of 200 feet and one of 250 feet respectively. By June 1st, 700 feet, in two sections, had been washed out by drift. The drivers were set to work, and by June 22d, had closed all gaps but one of 250 feet. Between July 16th and 23d, on four different occasions, 1,254 feet of dike in all was carried away. High current and deep water made repairs difficult, but by September 20th, the north dike had been restored to a length of 3,995 feet, and the south to a length of 959 feet, leaving a gap of 900 feet between them, as a new channel had not as yet scoured out enough for the purposes of navigation. By November 1st, 1883, the north dike was covered with sand for 1,850 feet, and partly covered for all but 500 feet. The main dikes had cost by this time $90,860.14, for 8,305 feet constructed.

After the first fall of the river in 1884, the water broke through the upper portion of the old main dike and was cutting a channel through the deposit at the foot of Osceola lower bar. To counteract this a dike called Osceola No. 4, was begun May 12th, and runs from the main dike to the Arkansas shore, a distance of 2,450 feet. The suddenly scoured channel, 600 feet wide and 30 feet deep; construction in water as deep as 51 feet, large quantities of drift; current very swift, as high as 8 or 10 miles per hour; all marked this work as difficult in the extreme.

The work was intermitted owing to low water in the fall of 1884, but completed in February, 1885. At last reports the dike is in good order, and the fills behind it large.

BULLERTON TOWHEAD PROTECTION.

The protection of this island on its head, partly along the west side, and for a long distance on the eastern side, has been a long and expensive operation. Begun in 1882, a mattress 1,428 feet long and 107 feet wide was sunk. In 1883 construction was carried on when a current of 5.35 miles per hour was observed, and the water from 25 to 50 feet deep, off the bank. Between December 1st, 1882, and November 1st, 1883, the following work had been done:

Mattress made and sunk 10,319 feet by 100; 355'x150'; 298'x75'; 394'x37'. Revetment made, 8,140'x70'; 4,985' loaded. At this time this protection had cost $157,087.61 for 21,095 squares of 100 feet each.

During the following year, to complete the work and repair damages, the following was done:

Foot mattress made and sunk....................1,423 squares.
Revetment made.....................................2,636   "
Revetment loaded....................................6,187   "

BULLERTON CROSS DIKES.

To close the channel between Bullerton Towhead and the Arkansas shore, a dike was begun, in 1883, at each end, and finished that year, to a length of 1,200 feet from the towhead, and 570 feet from the Arkansas shore. By December, 1884, this dike was closed entirely, the channel being established to the east of Bullerton Towhead. A second cross dike, known as No. 2, was begun July 5th, 1884, and was substantially built. It was finished without accident that fall. These dikes are in good condition, except for a break in March, 1885, of about 100 feet in No. 1, and large deposits are forming behind them. On October 1, 1884, the Bullerton Towhead revetment had cost $231,586.22 for 8,700 linear feet, and the two cross dikes had cost $135,408.18 for 3,200 linear feet. The Bullerton works as a whole had cost $460,641.27, or about 24 per cent. of those of the whole reach.

The record thus briefly sketched is that of an undertaking never before achieved. The main current and channel of the Mississippi River have been controlled and placed where desired, immense areas of land have been acquired, and the former main channel filled to a high level; the erosion and wash, the drift and floods, have been met; and slight, but successful, works of construction have, in less than three years, done what was expected of them, and are now holding their own. From Osceola to Bullerton Towhead the seven miles of the west bank of the river have absorbed about 44 per cent. of the expenditures upon the whole reach.

PLUM POINT DIKES.

To complete the contraction proposed, a line of main dike was proposed upon the eastern shore, to run parallel to the new line on the west; and with a gentle curve opposite the lower end of the lower Osceola bar, and the upper part of Bullerton Towhead, extend down stream and build up a new eastern bank behind it. Of this main dike, supports were made in six cross-dikes.

Construction began September 20th, 1883, and up to December 1st, 1883, had been built: main dike, 2,008 feet; and 1,437 linear feet of the first two cross-dikes. Upon March 10th, 1884, construction was resumed, and during 1884 continued without special accident, or unusual variation in methods. General figures of this work are as follows:

DIKES.	Lineal feet, Oct. 1, '84.	Foot Mat. lineal feet.	Tip Mat. lineal feet.	Grillage. lineal feet.	Bank pro-tect'n –feet.
Main Dike............	3,025	1,699	....	....	....
Cross-Dike No. 1...	985	....	....	....	....
" 2...	1,600	....	....	....	....
" 3...	2,100	....	1,800	300	220
" 4...	2,000	....	1,725	510	200
" 5...	2,200	....	516	....	....
" 6...	3,100	650	....	2,850	....

The channel way varies from 4,000 feet in width, at the foot of Bullerton Towhead, to 4,800 at the head. The dikes have served to close one channel, to the left of Bullerton bar, and have secured large deposits behind them. The Plum Point dikes had cost $339,152.61, for 15,010 linear feet, up to October 1st, 1884; being 17½ per cent. of the whole. Time will show if this system should be extended further down stream.

CRAIGHEAD POINT REVETMENT.

This work was recognized as a necessity to the plan of improvement of the reach, but became a matter of urgency in 1884, when the bar in front of the lower Osceola bar moved down stream, partially closing the entrance to the channel, immediately along the front of Bullerton Towhead, and the water, passing mainly around the head of Yankee bar,— about a mile lower down,— and through Bullerton bar, united to cause rapid caving, in the Arkansas bank, under the assault. On August 5th, a mat 150 feet wide was begun, and by the 27th, 1,197 feet were made and sunk; all but 100 feet, which were lost in sinking. On the 28th, a new mat 175 feet wide was begun, with a heavy boom of cypress piling. Thirty-three feet of mat were made on the barge before shifting, and September 2d, an attempt was made to launch it, but, owing to the strong current, the instant it struck the water the boom broke, and the mat doubled up and slid from the barge; but it was held and sunk near the shore. On the 3d, a new piece of mat, 150 feet wide, was begun; and on the 5th, it met the same fate. On the 9th, a new mat, 175 feet wide, was started, with two booms ten feet apart, and seven piling run up into the mat, which was commenced on the upper boom. On the 16th, 38 feet were successfully launched, and held by seven lines, and two wire ropes. Drift began to collect, almost imperceptibly, and on the 20th the two wire ropes parted, causing the head of the mat to sink about two feet. Drift was removed, where possible, and, by the 29th, when 581 feet had been built, and all was ready for sinking it, all the lines parted, (six 2-inch, seven 1½-inch, and one wire rope), and the mat, mat barge, and three other barges, were carried down the river. The barges were caught; the mat lost. Another mat, 150 feet wide, was begun; 400 feet built in October; and in November 660 feet were sunk, just below Bullerton Towhead. A collision, on November 6th, carried away the mattress boat, but the mattress held, because it had twenty cables running throughout its length, and was held by thirteen head, and ten side lines. Although 1757 feet of bank were mattressed by this time, the officer in charge thought it but a purchase of experience, nothing more. Caving of all kinds, above, below, and behind the revetment, was in progress, and when the season and work closed nothing satisfactory had been accomplished.

On October 1st, 1884, Craighead Point revetment had cost $29,106.45 for 1,050 feet.

The total expenditures upon Plum Point reach, on October 1st, 1884, amounted to $1,923,558.48, and on June 30th, 1885, amounted to $2,379,019.12.

GENERAL RESULTS.

The Commission say, as to this reach, that during the low water of 1884 there were five channels through Bullerton bar, and that in time the Bullerton Towhead channel, alongside the river side, finally deepened to 12 feet. More or less trouble is expected at this place till the contraction works can receive their full development and have time to act. The effect exerted by them in their present incomplete state leaves no doubt, however, as to the ultimate success of the work. Much revetment remains to be done, the dike work is nearly completed, and needs watching and repair. The officer in charge, Captain Knight, says: "The low water survey of 1885 and 1886 will show whether or not advance has been made in securing a satisfactory low water channel along the front of Bullerton Towhead. This season all that can be claimed for the dikes is that they assisted in removing obstacles to low water navigation earlier than nature, unaided, removed them elsewhere in the river. Next season I believe they will materially aid in preventing their recurrence."

In May, 1885, the officer in charge (Captain Leach) reports that the Bullerton region could be passed in three ways with 13½ feet water, and that a 12 foot depth had been maintained throughout.

NEW MADRID REACH.

No work has been done on this reach beyond the necessary surveys and the preparation of a plan and the purchase of plant. Forty barges, a machine-shop boat, twenty pile-drivers, and sundry material, were bought for the reach, at an expense of $193,550. This plant was all in use at Plum Point and elsewhere, excepting ten barges and ten pile drivers. The above works and the necessary surveys have included all belonging to the first district. The money disbursed within the district by October 1st, 1884, was $2,272,765.50.

CHAPTER XV.

In the second district the principal work, besides the protection of the water front of Memphis, has been the revetment of Hopefield Bend, on the Arkansas shore, just above Memphis. The project contemplated the protection of the bank for a distance of about two miles. Active operations began in December, 1882, and by October 31st, 1883, the total amount of work done was:

Bank grading.................................67,242 cubic yards.
Mattress, 140 feet wide made.............. 7,529 linear feet.
"   "   "   " sunk.............. 6,956   "   "
Upper bank protection made.............. 4,922 square yards.
Total revetment....................:.............120,918   "   "

. Mattresses were sunk in water from 40 to 70 feet deep.

During 1884 work was carried on under many adverse circumstances, and attended with many losses, caused by drift and currents. All of the subaqueous mattress, from where the upper bank revetment ended, down to the end, had disappeared after the flood, which left only 5,700 linear feet in fair shape. From November 1st, 1883, to October 1st, 1884, the following work was done:

Mattress 150 feet wide made................. 1,278 linear feet.
"   "   "   " sunk.....................1,115   "   "
"   140 "   " made...................5,539   "   "
"   "   "   " sunk...................4,895   "   "
"   "   "   " lost in sinking......1,260   "   "

Some repairs and extension were made during the next two months, and the total expenditures upon this work, known as Memphis Reach, was, upon March 31st, 1885, $469,352.31.

The levee work will be noticed later.

In the third district the Lake Providence Reach is the prominent stretch where construction has been done. This reach begins at Carolina Landing, 517 miles below Cairo. The river there makes a long sweeping bend, known as Louisiana bend, by Bunch's cut-off, to Opossum Point. Opposite here is a wide stretch, with large bars, between the main river and Duncansby and Skipwith Landings. Sweeping to the south the river is very wide at Island No. 93 and Mayersville, 533 miles below Cairo. Three and a-half miles lower is Baleshed bar, and at 540 is the head of Stack Island. Between these two the low water width is a mile and a-half. Lake Providence Landing, Louisiana, is 542½ miles below Cairo. The river is again very wide for five miles, terminating in Ajax Bar. Then, with a sweep to the east at Point Lookout and Island No. 95, and again to the south, the river narrows at 533 miles below Cairo, the end of the Lake Providence Reach.

The project for improvement contemplated the reduction to a width of 3,000 feet in general, by pile dikes, and the revetment of Bunch's or Louisiana bend, Island 93, Stack Island, the Louisiana shore above and below Lake Providence, and at Point Lookout.

The characteristics of the stretch are sharp bends and very deep water and rapidly caving banks at the head and foot of the reach; eight crossings from bank to bank, on some of which but four and a half feet of water was found at low water in former years; a stretch of 7 miles of straight river below island 93; excessive width in three places. Below Lake Providence the river is 9,600 feet wide at high water, and much obstructed by shifting sand bars of great extent.

PILCHER'S POINT REVETMENT.

The uppermost or Louisiana bend showed more rapid caving of the banks in 1881-3; than any part of the reach, producing effects felt for miles. For 7 miles on the Louisiana shore the banks caved so much that in two years the Mississippi low water shore line occupied in places, the position of the Louisiana shore line then; or in other words the river caved its entire low water width in two high water seasons, or about 1,500 feet. The water was very deep, reaching 100 feet at low water. In May, 1883, a party was organized to begin the work of revetment by clearing the banks. Low water and the sickly season retarded work, and only about half a mile of mattress was made, of which 1,228 linear feet were sunk by November 26th. Work was resumed in 1884, and mats 180, 168 and 165 feet wide were sunk, the longest mat being 3,440x180 feet. On June 1st, 1884, 1,690 feet of the work of 1883 were in place, and 2,308 feet had been carried away. The caving had been from 150 to 600 feet in width along this distance. Of the work done from November 1st, 1883, to October 1st, 1884, the following is the summary:

Bank graded..			11,421	linear ft.
Mattress made and sunk.................	150 ft. wide	1,030	"
" "	160 "	460	"
" "	168 "	1,779	"
" "	180 "	5,372	"
" "	165 "	4,195	"
Mattress sunk.............................	165 "	3,074	"
Grillage	50 "	1,768	"
Upper bank revetment ....60 and 80	"	5,725	"

During the next four months the mattress work done was as follows:

Oct. and Nov., 1884, made and sunk 182 ft. wide 1,888 ft. long

December,	"	"	35	"	300	"
"	"	"	50	"	115	"
"	"	"	100	"	882	"
"	"	"	125	"	164	"
"	"	"	188	"	1,531	"
January, 1885,		"	121	"	123	"

The total line of revetment was about equal to two and one-third miles. In May, 1885, it is reported that the caving in the completed work was about thirty per cent. of it.

THE DUNCANSBY SYSTEM.

The first works of contraction on the reach were built opposite Opossum Point, where the river was divided into two channels, by two tow-heads, or dry sand bars; the channel on the west being the deeper. It was intended to close the chute on the Mississippi side of the tow-heads; and at the head of the upper one, and in the upper chute, were built three cross-dikes, with foot mats and screens; and two, lower down, with screens, only, in 1883, to strengthen and connect with a pile dike, built in 1882, from the upper to the lower tow-head.

1882. Longitudinal dike, 6,805 linear ft.
1883. Main dike, woven mat, 1,615'x 60'; along piling.
" Cross-dike No. 1.... 100 linear ft.
" " " 3... 150 "
" " " 5... 545 " and 65 ft. foot mat.
" " " 6...2,105 "
" " " 7...2,061 " " 826 "
" " " 8...2,310 " " 1,650 "

Also 142,022 square feet shore mat, along main dike, and 3,097 linear feet of wattling.

By the caving in Pilcher's bend, in 1883, the attack of the current, on the Mississippi side, was changed to the head of this system, and parts of the upper dikes were washed out. A marked and decided fill had taken place behind the remains of the dikes. Along the longitudinal dike, sand was deposited as high as the top of the piling, in a short while. To meet the threatened flanking of the works, by the head, new dikes were built, three in number; and a dike across the head of the upper tow-head; which was also mattressed. These dikes were all held in 1884, and their object, the closing of the Skipwith Chute, was secured; though their future is problematical, and dependent upon the holding of the Pilcher's Point revetment, and the consequent permanence of the river just above the bend.

MAYERSVILLE, OR ISLAND NO. 93.

In 1882, were built 1,700 feet of hurdle mattress, along the front of this island; which was extended, by February, 1883, until about 1½ miles were revetted. Floods then carried away 225 feet, and damaged more. Repairs, reconstruction and extension followed; and during 1884 the following was done:

Woven mattress		45 ft. wide,	403 ft. long.		Washed out	
"	"	71 "	407 "		"	
"	"	100 "	1,774 "	525 ft.	"	
"	"	110 "	904 "		"	
"	"	125 "	310 "	Repairs.		
"	"	135 "	2,829 "			
"	"	151 "	2,906 "			
Grillage	" 32 & 50 "	2,172 "		780 ft. washed out		
Revetment................12,313 "		5,029 "				

Woven mattress and revetment remaining in good condition 5,735 ft.

Repairs have been made up to December, 1884, and since then some caving has occurred. It is considered important to hold this island if the river below it, is to approach a permanent condition.

A dike 2,300 feet long crossing the Mayersville chute at the head, and a cross-dike 805 feet long, one-third of the way down the chute, were built in 1882 and 1883. These dikes stood well until April, 1884, when the upper one was breached, but was repaired. The channel afterwards encroached upon the head of the island, carrying away all but 600 feet of the longitudinal dike.

Directly opposite island 93, a series of dikes known as the Cotton-wood dikes, was built in 1884 to contract the channel way and hold it in the bend of Mayersville Island. A feature in these dikes is a parallel, interior longitudinal dike, in addition to the usual cross-dikes. Their construction resulted in a considerable fill below and behind them, and the whole is the most compact and symmetrical piece of dike work on the reach. Amount of construction in 1884.

Woven mattress...........................	4,602	linear feet.
Revetment........................	120	"
Main outer dike...........................	4,471	"
" inner "	1,350	"
Cross-dike No. 1...........................	863	"
" 2...........................	1,285	"
" 3...........................	558	"

THE BALESHED DIKES.

To contract the river to the proposed width in the long stretch between Island 93 and Stack Island, a system of main longitudinal dikes was projected along down the Baleshed bars, and up from Stack Island, the whole to be the new eastern shore for a distance of about

4⅔ miles. Cross dikes were also intended, thirteen in number, to connect the main dikes with the Mississippi shore, the whole system to secure deposits in, and close the deep pools of the Baleshed chute and the Stack Island chute, and thus collect the water then flowing down these chutes, into one channel flowing to the west of these works. Of this work, a part of the long dike at the head of Baleshed chute, and parts of four cross dikes had been done during the low water of 1882, to the level of 17 feet below low water mark, the whole work done being as follows:

Single row pile dike...... .... 4,289 linear feet.
Double row "   " 18,995   "   "
Screens built............................... 5,375   "   "
Hurdle wattling.............................   395   "   "
Foot mats, 30 feet wide..................... 7,379   "   "

In February, 1883, work was resumed, and was prosecuted continuously until the ensuing high water in the following winter. The channel behind Baleshed bar was enlarging, and large amounts of drift running when work began, and many piles were washed out as soon as driven, but a lodgment was at last secured, and by September, 1883, 28,000 linear feet of pile dike had been constructed, and a great fill had been secured. Of twelve cross dikes six had been built by that time, from the shore to the dike, and six were incomplete. The three upper cross dikes had foot mats. The high dikes all reached to the 25-foot stage, and were wattled for six feet in height above the bar surface. In front of the main dike a foot mattress ranging from 60 to 100 feet in width, was laid in places, and in others, grillage mats. The fill caused by these dikes was enormous, and the channel was deflected towards the Louisiana shore.

In 1884 a high water dike was built inside of and parallel to the main dike from cross dike No. 1 to No. 4, and on the outside of the main dike to cross dike No. 6, in all, about a mile, and repairs were made where necessary. The head of the system had caused deposits by this time, which were dry at half stage.

Completion of the cross dikes while the water permitted, and finishing wattling work, were the final operations on these dikes in December, 1884, and January, 1885. No injuries have been reported to this system of dikes up to July 1st, 1885.

BALESHED SYSTEM OF DIKES.

DIKES.	Pile Work. lineal feet.			Wattling. lineal foot.		Foot Mat sunk. lineal feet.		Woven Mat along piling. lineal feet.	
	1882.	1883.	1884.	1883.	1884.	1883.	1884.	1883.	1884.
Main Dike	6,665	13,733	1,671	1,708	1,865	3,239	1,915	6,741	2,467
Cross Dike No. 1	......	933	......	995	......	933	......	......	......
" " 2	1,096	1,195	......	......	......	810	......	......	......
" " 3	......	1,172	......	......	872	700	672	......	300
" " 4	1,575	......	......	......	......	717	.. ..	......	......
" " 4 new	......	661	......	......	......	......	......	......	......
" " 5	1,676	......	......	......	......	581	859	......	......
" " 5 new	......	983	......	......	......	1,090	......	......	......
" " 6	......	1,452	......	......	400	1,749	....	......	......
" " 7	......	1,204	......	......	915	1,125	75	......	......
" " 8	......	1,011	......	......	......	......	......	......	......
" " 9	......	1,097	......	......	......	......	......	......	......
" " 10	......	924	......	......	......	......	......	......	......
" " 11	1,507	894	......	......	......	......	......	......	......
" " 12	......	583	......	......	......	918	......	......	......
" " 13	1,233	......	......	......	......	......	......	......	......

The dike from Stack Island up stream was built in 1883, to a length of 5,250 feet, and of this 1,429 feet was washed out, and replaced during construction. The chute which this dike would close reached a depth, in places, of 83 feet at low water. Along the main dike 3,338 lineal feet (102,725 square feet) of foot mat were sunk. In 1884 a dike was run from the main dike to the Mississippi shore, and the main dike strengthened by a parallel construction; amounts as follows:

DIKES.	Length. feet.	Foot Mat. feet.	Foot Mat. square feet.
Main Dike	2,249	2,369	95,760
Dike No. 1	2,636	2,139	88,895

At high water these dikes caused a large deposit, closed the secondary channel, and deflected the main channel to the right of the island as projected—removing from its path immense deposits of sand.

A chute, called Elton chute, which was 30 feet deep at low water, on the Louisiana shore, was closed in 1883 by a series of six short spur dikes, as a part of the contraction here.

DIKES.	Length. feet.	Foot Mat. linear feet.	Foot Mat. square feet.
Main Dike..	943	1,475	44,250
Cross-Dike No. 1..........	746	....	....
" 2...............................	887	950	26,600
" 3...............................	975	975	29,250
" 4...............................	300	....	....
" 5...............................	484	....	....
" 6...............................	435	....	....

The Stack Island and Elton dikes are in good order.

LAKE PROVIDENCE REACH, SOME GENERAL STATISTICS.

DIKE, MATRESS WORK, Etc.	1882.	1883.	1884.
Total Dike Work—constructed during year, linear feet.....	22,284	44,235	27,165
double row pile dike, " "	18,995	25,187	5,463
three rows " "	......	14,150	20,841
four " " "	......	565	861
five " " "	......	1,997	......
washed out during year, . "	......	6,432	8,294
Mattress Work—foot mat bet. piling sunk, "	7,739	22,141	20,594
woven mat along piling, "	5,375	10,576	6,196
wattling, "	395	5,800	9,948
shore mat made and rocked, square feet...	......	142,022	83,600
Bank Revetment Mattress—in good condition, linear feet...	1,719	5,220	......
washed away, "	......	8,505	......
made, "	......	......	10,928
Revetment—in good condition, "	......	1,700	......
washed away, "	......	425	......
made, "	......	......	15,538

Total amount of pile dikes standing October 1st, 1884, 63,106 linear feet.

Total expended .March 3d, 1881, to June 3d, 1885, $2,240,285.73.

During 1884, no water less than 15 feet in depth, has been found along this reach in the channel way; and in 1885, not less than 14 feet; and the Mississippi River Commission repeat in their report for 1884, the statement as to favorable results secured, made the preceding year.

MOUTH OF RED RIVER.

The waters of Red River flow, in part, into the Mississippi River, and partly into the Atchafalaya, and thence into the Gulf. The relative amounts flowing through these two channels, at high and low water, depend upon the relative depths and widths of the two. At high water of the Mississippi, a direct flow to the Gulf is established, down the Atchafalaya, of part of its waters; and by overflow of the banks below Red River, of more, through the same outlet. For many years the relative conditions have been slowly changing, by the enlargement of the head of Atchafalaya, and the silting up of the mouth of Red River. By December, 1876, this silting had so far progressed that steamers could not enter Red River from the Mississippi. In the act of 1878, $150,000 was appropriated for the improvement of the mouth of Red River; and $40,000 in 1879. This was expended in furnishing a temporary relief, by dredging a channel through the bar at the mouth, and in surveys, and studies for a plan of permanent improvement. Having completed such, the officer in charge, Major Benyaurd, proposed to abandon the mouth of the river, and establish communication with Red River by a lock and canal, from the Mississippi to Bayou Plaquemine, and thence into the Atchafalaya, and up it, to Red River. This plan was submitted to the Board of Engineers, consisting of Colonels Tower and Newton, and Major Abbot, who reported [1]April 20th, 1880. Giving an account of the main facts, it was shown that the mutual relations between the three rivers were of great importance in time of flood, the Atchafalaya furnishing an outlet for the floods of the two upper rivers, which could not safely be interfered with, and that for the present it would be best to continue the dredging. Major Benyaurd's plan was thought to be more expensive than estimated. A study of the rate of enlargement of the upper Atchafalaya was recommended.

The preliminary report of the Mississippi River Commission postponed the consideration of this question, until the matured views of Major Benyaurd were presented.

Dredging was continued, without any permanent results, and with very little temporary relief; an entire cessation of navigation occurring for a short time in 1881, and the caving, and settlement into the channel of such large masses of the banks, being so constant as to forbid any hope of success. The Atchafalaya continued to enlarge. The report of the Mississippi River Commission, dated November 25th, 1881, states that the capacity of discharge of the Atchafalaya was nearly equal to Red River, which now discharged but little into the Mississippi. The outlet from the Mississippi to the Atchafalaya was almost constant, and at times very large, and the tendency is to increase; and the plan proposed, was to check future enlargement of these channels, by a broad, continuous brush sill across the outlet from the Mississippi into the Atchafalaya; and thorough protection of the banks. Further surveys were ordered. Dredging was continued, but again failed, in the low water season of 1883, to maintain a channel; in fact, for a time, the water way entirely disappeared. Surveys showed that the sectional area of Old River, the former course of Red River, was decreasing, and that of the Atchafalaya increasing.

In the report for 1884, the Commission made a report of a plan for the rectification of Red and Atchafalaya Rivers, and discussed, in extenso, all proposed plans, also.

---

1. 1880, II, p. 1284.

11

A system of brush dams was proposed for the Atchafalaya. These, built on a broad foundation of brush, were to be of concentric, and diminishing cribs, and with crests just below low water, with a channel depression of five feet. From the levees on either side of the river, and opposite the ends of the dams, earthen spurs should be built out as far as possible. These spurs and all adjacent surfaces of levees or banks should be substantially rip-rapped, if necessary. As this plan would cause a large increase of flood discharge in the Mississippi below Red River, the levees would be raised, of necessity, along that part of the river, and the cost of this was also included. Total estimate, $4,800,800.

After the Commission took charge of this work there was expended at the mouth of Red River to June 30th, 1885, the sum of $98,386.32.

NEW ORLEANS HARBOR.

A Board of Engineers, consisting of Majors Weitzel, Benyaurd and Howell, of the corps of engineers, and Messrs. Harrod, D'Hernecourt and Wood, assembled at the request of the City Council of New Orleans, reported April 8th, 1878, [2] on the dangers threatening the harbor of that city on account of caving banks, and the remedies to be adopted. The surveys had shown that caving occurred, more or less, along more than ten miles of bank. The plan proposed was to protect by a revetment of brush and stone some 18,700 lineal feet in all, and to build a timber bulkhead along 7,500 feet of the main front; the whole work estimated to cost $476,000. An appropriation of $50,000 was made in 1878. Work was begun in revetment, and 2,002 lineal feet were constructed of mattresses, 24 feet wide, and sunk with a lap of ten feet. The total area covered was 326,650 square feet. Operations were carried on during the successive years under two different contracts in execution of the plan proposed. But from a combination of circumstances; the difficulty of operating near the shipping, the difference between the protection of cheaply built wharves, and that of a sloping bank; inefficient methods of construction; and lastly the insufficiency of the revetment already laid; all work was summarily abandoned in the fall of 1881. The subject was referred to a Board of Engineers consisting of Majors Suter, Benyaurd and Stickney, and Captain Ernst, who reported October 5th, 1881, [3] that the present plan of improvement, whether viewed as to its general merits or its details, should not be continued. A second report from the same board, dated February 20th, 1882, proposed that the problem of wharf protection would be best met by proper wharf construction, and indicated the methods; and that to secure the bank in Carrollton bend a single mattress, 10,000 feet long, if practicable, and about 400 feet wide, should be laid, at a cost of about $280,000; but that the success of all work depended upon the ability to construct without interrup-

2. 1878, I., p. 615.   3. II, 1882, p. 1359.

tion from shipping what was intended as a protection, in either locality.

This work was turned over to the Commission by the act of 1882, who adopted the plan for the revetment of Carrollton Bend. Work began on this November 9th, 1883. When 470 feet had been laid, the rising river forced a suspension of work. Later on, an extensive caving in Gouldsboro bend made it more desirable to protect that, and the plant was transferred there, and a portion of the fall of 1884 was spent in building out spurs in that bend, until the funds on hand were exhausted. In four appropriations $260,000 have been given to New Orleans harbor.

LEVEES.

In the preliminary report of the Mississippi River Commission it is stated that there is no doubt that the levees exert a direct action in deepening the channel, and enlarging the bed of the river during those periods of high water, when by preventing the dispersion of the flood-waters over the adjacent low lands, they actually cause the water to rise to a higher level within the river bed than it would attain if not thus restrained. Surveys give some evidence that up to 1858 the agency of a perfecting system of levees, secured a river bed .of greater area of mean cross section, of greater enlargement, and of corresponding deepening. During the ensuing twenty years, it is known that the levee system was interrupted by a great number of crevasses between Cairo and Red River. It would appear from known facts that a closure of these crevasses might be expected to accelerate the removal of shoals caused by them, and together with contraction works might effect a lowering of the flood level, which will in time render the maintenance of these levees above Red River unnecessary. The breaks in the levees are estimated equal to about 8,065,700 cubic yards; estimate for repairs, $2,020,000. Two members of the Commission dissented from the views in a published report.

In the report of the Commission for 1881, many surveys were made and facts noted upon the levee question, and collected together, and the deduction was drawn that they corroborated the views expressed by the large majority of the commission in the former report, and repeated in this.

In the report for 1882 the Commission announce as the result of studies and surveys made by them, that as a part of a complete system of channel improvement, levees should be built, and in conformity with these general points. The standard of elevation for levees should be sufficient to confine floods most frequently recurring. The restraint of great floods, which recur only at long intervals, would involve too great cost. The system should be generally one of restraint in the interest of navigation, the extent to be determined by considerations of economy. At outlets and depressions of the bank the levees are the more important and the more costly, and should be made secure, by giving the levees great width and height. To determine the suitable elevations will require careful study of the surveys, and especially of flood heights. Of the appropriation of $4,123,000, made August 2d, 1882, the Commission considered that the sum of $1,300,000 could be judiciously expended in closing existing breaks in the levee. It was deemed expedient to apply the sum first upon the lowest parts of the river flowing along the fronts of the great basins, and thence upward as far as the allotment would extend. New breaks were to be repaired first, and as long a line of levee as possible, rather than short parts, to a greater height. A board of the executive engineer officers of the river districts was authorized to make the distribution of the allotment, which was as follows:

Atchafalaya basin..$110,000
Tensas basin.. 750,000
Yazoo basin.. 440,000
Bonnet Carre crevasse................................ 15,000

Contracts were then made, and put in execution that year, for the repairs of levees, as follows: In the second district, on the Yazoo front, four contracts amounting to 291,500 cubic yards, at prices running from 22 to 32 cents per yard; in the third district, nine contracts along the Tensas front, amounting to 1,338,000 cubic yards, at from 18½ to 25 cents per yard; and fourteen contracts along the Yazoo front, amounting to 1,380,000 cubic yards, at from 18 to 29 cents per yard; and in the fourth district, eleven contracts along the Tensas front, covering 51 miles of levee, and 1,764,291 cubic yards. All of these contracts were carried out during 1882 and 1883.

In the report for 1883 the Commission recommended the continuation of levee work on the fronts of the Tensas and Yazoo basins during the ensuing year; and after discussing fully the results of the preceding year's surveys and constructions, give the same general conclusions as to the desirability of repairing the breaks in existing levees, as already adopted by them.

In the report for 1884, after carefully reviewing the reasons for forming the conclusion it is said:

"We therefore conclude that levees, such as have been described, are, in connection with an equalization of width, and the prevention of caving, an important part of any general and systematic plan for the improvement of the navigation, and the prevention of destructive floods; and we do recommend the construction of new, and raising of existing levees along all parts of the river, where the highlands are too remote to check the passage of large volumes of flood water, outside the bed of the river; or, in other words, on the entire right bank, and on the left bank below Baton Rouge, and from the Yazoo River to Horn Lake, below Memphis."

In the completion of the levee system, it was proposed to enclose each bottom, the Manchac, LaFourche, and Atchafalaya, constituting lower Louisiana; the Tensas, Yazoo, White, and Saint Francis, in order, from below, upwards. In each bottom, as it is reached in the preceding order, the most important levees, as restraining the greatest escape of water, should be first constructed.

From the appropriation of 1884, the Commission allotted $175,000 for the repair and preservation of levees; and this was distributed by the board of executive officers, as follows: In the second district, $15,000 to Long Lake levee, Arkansas; in the third district, $20,000 to the Yazoo front, and $50,000 to the Tensas front; in the fourth district, $90,000 to the Tensas front. These works were put under contract, and, since then, have been executed. Besides the sums referred to, special allotments have been made for special levees, at different times; and, up to June 30th, 1885, the total expended in the repair, and construction, of levees, is as follows:

Second District,	Long Lake Levee.........$	15,000
	Yazoo Front................	80,950
	Opossum Fork	25,000
Third District,	Yazoo Front..................	364,875
	Tensas Front..................	411,108
	Bonnet Carre Crevasse.....	15,000
Fourth District	Tensas Front.................	548,258
	Atchafalaya Front..........	133,504

Total levees.................................... . $1,593,695

This is a little less than twenty-one per cent. of the total expended in construction below Cairo, under the Commission.

The views of the Commission, upon the subject of levees, have never been concurred in by all of its members, but dissenting reports have been submitted upon all the theories upheld by the majority.

It has not been my intention to attempt to summarize the arguments upon the question of the influence of levees upon the navigation of the river. It has never been doubted that the work done upon the levees has been of advantage to the interests of navigation, and to the occupants of the protected territory; but it is disputed that the levees will exercise a beneficial effect upon the low water channel of the river, or in lowering the flood line; both of which results are claimed as certain to follow, by the majority of the Commission.

The literature upon this subject is extensive; the surveys, and information presented, are minute and extensive; and a student will draw his own conclusions. I have endeavored to present here, very briefly, the development of the conclusion of the majority of the Commission, as to levees, in a spirit of fairness, and to trace the actual construction, under the plans adopted, as a part of the "practice" of the improvement of the lower Mississippi.

CHAPTER XVI.

THE ILLINOIS RIVER; FOX RIVER, WISCONSIN; COOSA RIVER, ALABAMA; AND RED RIVER OF THE NORTH.

In accordance with the act of June 23d, 1866, a survey was made of the Illinois River, from Grafton to LaSalle, 224 miles, by Lieut.-Col. J. H. Wilson. He proposed as a result of this to improve the river by a system of locks and dams, believing that a sufficient depth could not otherwise be procured. Estimate, $3,123,796. He also proposed to enlarge the Illinois and Michigan Canal so as to adapt it to the use of the largest boats plying on the Mississippi River; estimate, $18,250,110. All locks were to be 350 feet long, and 75 feet wide, and the canal and river to have a navigable depth of 7 feet. [1] The proposed locks and dams were as follows:

LOCKS.	Lift. feet.	Height of dam. feet.	Length of dam. feet.	Length of pool. miles.	Estimated cost.
Lock No. 1, Six Mile Island.	4.	11.0	1,440	25½	$ 491,449
" 2, Columbiana......	4.	8.0	1,240	28	459,746
" 3, Naples............	5.	9.5	880	25¼	405,746
" 4, Frederick.........	5.8	10.3	980	43½	420,746
" 5, Spring Lake......	5.7	10.8	800	49¼	394,622
" 6, Chillicothe.......	7.5	14.0	930	42⅓	415,688

The total fall from LaSalle to Grafton is 29.2 feet, and the average fall per mile varying only slightly from 0′.4; the maximum was 0′.4 per mile at Beardstown bar; the minimum 0′.01 in Lake Peoria.

Flowing with a sluggish current, the river wanders through a valley of swampy land, from one to six miles wide. The banks are low, rising from three to eight feet above the river surface at medium stage. The general course of the river is noticeably direct; the straight reaches are almost invariably deep with a muddy bottom; the shallows occur at the bends, at confluent channels and the mouths of creeks. The amount of material brought down by floods is small. In the long reaches the depth found was from eighteen to thirty feet. For fifteen miles below Peoria the river expands to seven or eight times its ordinary width. Low water occurs in July, August and September, and navigation is virtually suspended, and for five months in the year ice causes a suspension of navigation. The low

---

1. Ex. Doc., No. 16, 40 Cong., 2st Session, H. R.

water discharge at Hennepin was found to be 633 cubic feet per second. [2] The highest water known at LaSalle in 1883, was about 28 feet above low water mark. Surveys of the Illinois continued in 1867, and the matter was referred to a board of engineers, who reported December 17th, 1867, that "the only feasible route for deep water communication between the Great Lakes and the Mississippi River, equally adapted to military, naval and commercial purposes, was by the line of the Illinois River, and the Illinois and Michigan Canal."

The board proposed five locks and dams, at Henry, Copperas Creek, LaGrange, Bedford, and Six Mile Island, at an estimated cost of $1,770,000, and expressed doubt whether a navigable channel could be secured by dredging and wing dams. [3] In 1869 the State of Illinois directed the building of a lock and dam at Henry, as a first step in the improvement; and operations were carried on by the state in substantial conformity to the plans of improvement recommended by the board. From the appropriation of 1869, $84,150 was allotted for improving the Illinois River. In 1870 Colonel Macomb, then in charge, carried out the modified project prepared in 1869 by his predecessor, and began a system of auxiliary wing dams, together with dredging, hoping this would be in the interests of navigation, since the funds on hand would not allow the building of locks and dams. From 1870 to 1877 operations were continued by dredging, and building wing dams at points where thought best.

In the meantime the state had built a fine lock and dam at Henry, while the government had prepared the river bed from that place to Copperas Creek, in the vicinity of which the second lock and dam were to be built. In 1873, in accordance with some arrangement made, the government began the foundations of this Copperas Creek lock, which were completed in September, 1874, then turned it over to the state, which finished the lock. Dredging was continued to the end of the fiscal year, June 30th, 1878, when it was suspended. Captain Lydecker relieved Colonel Macomb in 1877. At the end of this period of operations it may be summed up that the state works at Henry and Copperas Creek had secured a reliable 7-foot navigation for a distance of about 100 miles. On that part of the river below Copperas Creek a navigable depth of four feet had been secured over some bars, yet others, having only two and a-half feet, limited navigation to that depth.

---

2. 1868, p. 448, p. 437.   3. 1879, II., p. 1572.

Total dredging, by contract, 811,434 cubic yards...$235,785
Time work by dredge, snagging, etc., 1,655 hours... 17,051
Brush and stone dams, 6,000 linear feet.............. 29,117
Foundation Copperas Creek lock..................... 62,359
Removing wreck near Peru........................... 450

Total paid to contractors..........................$344,762

Aggregate length of dredged channels, 123,320 linear feet; 40 to 150 feet wide; material used in forming dikes and dams.

Total amount appropriated before 1878...... ....$474,150

Captain Lydecker urged the necessity of extending the slack water system, by adequate appropriations. The Chief of Engineers agreed with these views, but decided that the intention of Congress was to continue the use of dredging and wing dams, and the Secretary of War coincided in this decision.

Accordingly, a dredge was built, a tow boat, and dump scows bought, and work was continued upon this plan, during 1879, 1880, and 1881. In that time 199,738 cubic yards were dredged, and 5,858 linear feet of dams built, consuming 254,106 cubic feet of material. Appropriated $150,000 in 1879, and 1880. The act of March 3d, 1881, gave $250,000, with which, as an adequate sum for a commencement, it was decided to begin a new lock near Kampsville, opposite Columbiana. [4]This would correspond in size with those already built on the river; 350'x 75' between gates; walls of head bay extend 70' above the upper, and of the tail bay 50', below the lower hollow quoins. The lift of the Kampsville lock will be 7.2 feet, (and of the LaGrange lock 7.4 feet), the foundation consists of a system of piles, on which is secured a grillage and platform, covering the entire area included between the exterior lines of the lock walls. Five rows of sheet piles cross the foundation area; one at the head, one at each mitre sill, one at the foot, and one in continuation of the sheet piling of the dam. The tops of the lock walls are to be 7 feet above the crests of the dams, boats passing over the dams when the water rises to that level. The walls are about 22 feet high, with vertical faces of cut stone, and a backing of concrete, with stone imbedded in it; the gates of timber. The dams will be of timber crib work, prolonged down stream so as to form an apron, filled with stone; protected against undermining by sheet piling, reinforced below by a mass of rip-rap, and backed above by a filling of clay and gravel; one end of the dam to connect, water tight, with the river wall of the lock, the other resting against a masonry abutment. Dredged channels are to be made through the bars at the upper end of the reaches, to give a low water depth of 7 feet. Contracts were let, in 1881, for material and machinery to begin construction, and the site of the lock pit was dredged.

---

4. 1881, III, p. 2176.

Extreme and continued high water, in the spring of 1882, prevented operations. Appropriated in 1882, $175,000. Major Benyaurd relieved Major Lydecker from the charge of the river.

Dredging for the lock pit at LaGrange began July 19th; and the coffer dam was begun August 23d, and finished October 4th. It was 700 feet long, by 200 feet wide. It was pumped out, and pile driving for the foundation began in November, and continued until January 6th, 1883. Operations were resumed July 13th, and continued to November 29th, 1883. The foundation was completed and the lock walls begun.

Dredging for the lock pit at Kampsville began September 1st, 1882; the coffer dam was finished September 22d, and the foundation begun October 14th. Work ceased December 23d. Work was resumed July 18th, 1883, re-enforcing the coffer dam, and dredging out deposits within it. The lock foundation was finished November 12th, and funds were exhausted.

The appropriation of $100,000 of 1884, was to be expended at LaGrange lock, in raising the lock walls, purchasing material for, and constructing the gates, which it was hoped could be finished therewith. The amount required for completion of the existing project is $447-·150.56; and the river cannot be used to advantage until the locks and dams are all built.

The principal commerce by river, from the Illinois, is connected with Saint Louis, whence the following statistics are obtained:

STEAMERS ARRIVING AT SAINT LOUIS FROM THE ILLINOIS RIVER, AND FREIGHT HANDLED.

YEAR.	Steamers. arrivals	FREIGHT.	
		Received. tons	Shipped. tons
1878...	263	124,785	18,300
1879...	234	106,620	9,140
1880...	260	155,605	9,935
1881...	205	....	....
1882...	214	....	....
1883...	163	94,205	4,715

[5]A survey was made, in 1882, for an enlargement of the Illinois and Michigan Canal, without change of location of canal trunk, or locks, with a view to the increase of depth from 6 to 7 feet; and in size of locks from 110'x 18' to 170'x 30'. The estimate, based on this survey, was $2,298,919. There are now 16 locks, covering a descent of 141.5 feet, in all. The State of Illinois has offered to the United States this canal, if intended for enlargement.

5. 1883, II, p. 1761.

If a water communication between the Mississippi Valley and the Great Lakes be desirable, for military, naval, or commercial, purposes,— and it is difficult to form arguments against such a connection,— then the Illinois and Michigan Canal, properly enlarged, and the Illinois River properly locked and damned, will be the most natural, direct, and economical line. The relations of the State with the line, and its willingness to further the project; the constant reports of the engineers who examine the subject; the action of the United States, its delays and limited appropriations, reproduce the same historical picture as seen in the case of the Louisville and Portland Canal; the Sault Sainte Marie Canal; the Muscle Shoals Canal; and the Des Moines Rapids Canal. An ample appropriation, an energetic construction, such as would be expected in the building of a great railroad, would rapidly solve the problem, presented eighteen years ago, and repeated annually since then.

FOX RIVER, WISCONSIN.

[6] The Fox River, Wisconsin, which with the Wisconsin River forms a water line (by canal) between the Mississippi and Lake Michigan, was improved by private enterprise through locks and dams. When the government purchased in 1872 the property of the canal company, the works were in such a dilapidated condition that navigation was practically suspended, and it became necessary to rebuild nearly all the locks and dams, and to dredge the channel to a great extent. This has been done in part as funds permitted, since that time, and in 1884 three feet depth of navigation was maintained upon the Upper Fox, and five feet upon the lower river, except at one place, where only four and a-half were found.

Besides current and extensive repairs on many old locks and dams, and extensive dredging of the channel, the following is a list of the principal works upon the lower river. There are four steamboats running regularly upon this part of the river, besides transient boats and tugs with scows. They transport principally lumber, tanbark, salt, oil, sand and bricks.

LOWER FOX.

DePere Lock, (old), thoroughly repaired.
DePere Dam, new.
Little Kaukana Dam, rew.
Rapid Croche Dam, new.
Kaukana Fourth Lock, new.
Kaukana Third Lock, new.
Kaukana First Lock, new.
Little Chute Combined Lock, new.
Little Chute Second Lock, new.
Little Chute Dam, new.
Cedars Dam, new.
Appleton Second Lock and Upper Dam, new.

---

6.   1884, III., p. 1929. (See page 107 ante.)

UPON THE UPPER FOX.

Grand River Lock is a new stone lock.

Princeton, White River, Berlin and Eureka Locks, are also new stone locks.

Eureka Dam is a timber dam with stone piers, the whole resting upon a pile foundation. It has a navigable pass, 50 feet wide at the north, and which becomes a sluice way during high water. On the Upper Fox are four steamers running exclusively upon it, besides occasional trips made by other boats. Three tow barges for freight.

The following table shows the tonnage locked through the various locks during a space of three years, and also the greatest amount of lumber rafted through any one lock:

		Tonnage.			Greatest amount of lumber rafted. B. M.
Lock No. 1, DePere, to Lock No. 19, Eureka	1880	18,703 to 32,513			582,000
	1881	15,338 " 30,404			604,000
	1882	17,398 " 40,190			482,000
Lock No. 20, Berlin, to Lock No. 26, Fort Winnebago	1880	2,680 " 5,865			19,000
	1881	3,249 " 3,950			1,956,000
	1882	2,404 " 2,762			40,000
Lock No. 27, Portage City	1880	852			1,562,614
	1881	494			4,152,998
	1882	890			285,000

The total amount expended for construction and maintenance since the purchase to June 30th, 1883, $1,646,140.26. Appropriated 1884, $160,000. For maintenance and necessary repairs $45,000 are needed annually. Major Houston was in charge until 1882, and since then Major Benyaurd, except temporarily, when relieved by Captain Marshall.

COOSA RIVER—ALABAMA AND GEORGIA.

[7] In 1875 an examination of this river was made from Rome, Georgia, to Gadsden, Alabama, 135 miles, and a plan of improvement proposed to secure a depth of four feet through a channel 80 feet wide. In 1876 an appropriation was made, and a systematic plan of action begun. Active operations have been carried on each year. Below Greensport is a succession of very broad and shallow reefs, extending over five miles of the river, and having a total fall of 24 feet, the greater part of which is confined to a length of about two miles. This part of the river requires a more costly work than will be necessary on any other part of the river included within the present project. The plans for improving these shoals are as follows: First, a longitudinal dam, 2,000 feet long, forming a canal, with a lock at the lower end. Second, a dam 1,000 feet long across the river, terminated at

---

7. 1883, II., p. 1501.

one bank by another lock. Third, the utilization of another chute as part of the canal, placing a dam and a third lock at the lower end. The locks are 210'x40' inside. At the close of the fiscal year 1884 this work stood thus: Lock No. 1, masonry and upper gates completed; lower gates ready to be placed; longitudinal dam finished. Lock No. 2, masonry completed, and ready for the gates; the cross dam and a short closing dam finished. Lock No. 3. All the stone required for this lock has been cut. Coffer dam inclosing the lock pit has been constructed, and the foundation excavated. With the funds on hand this lock has probably been pushed towards completion.

Five steamers ply on this river, carrying lumber, iron and cotton, and general merchandise between Rome, Georgia, and Greensport, Alabama. The total appropriated has been $418,700. Major W. R. King is, and has been, the officer in charge. When the stage of water is high, steamers pass over the rapids without difficulty.

RED RIVER OF THE NORTH.

[8] This river from the junction of the Bois de Sioux and Otter Tail Rivers, on the western boundary of Minnesota, flows nearly due north 197 miles, to the northern boundary of the United States. The course of the river in the same limits is more than twice as long. The current, except at the rapids is hardly one mile an hour. It flows through a very flat prairie, between clay banks from 20 to 60 feet high. The area drained within the United States is about 32,000 square miles; but the rainfall is small, between 13 and 20 inches, as measured at different points.

LOCALITIES.	Width of River.—ft.	Distance. miles.	Fall. feet.
Moorhead..................................	100	....	....
Goose Rapids, head......................	....	98.0	50'.66
" foot........................	....	98.9	55'.26
Frog Point...............................	160	120.0	71'.83

Difference between high and low water marks at Moorhead 36', at the boundary line 45'. Upon the settlement of this country this river has had a decidedly beneficial effect. The principal obstructions to navigation have been removed by dredging, and from 1879 to 1882, a channel with an assured depth of 3 feet at low water was secured for 80 miles below Moorhead, where but 1½ feet existed before. Below Grand Forks 18 miles of a 4-foot channel are now found, where were only 2 feet before improvement.

The clay of the river bed and the gentle current have made dredging more than usually successful, and permanent in effect. There were three steamboat lines operating on the river in 1883. [9] Total freight carried during five years:

8. 1874, I, p. 295.   9. 1884, III, p. 1610.

1879...35,718,731 pounds.
1880...43,301,515   "
1881...53,114,861   "
1882...63,303,629   "
1883...50,727,951   "

It has been proposed to build a lock and dam at Goose Rapids to overcome this, the main obstruction of the river, but work has not yet been begun, although $50,000 were appropriated. Between 1876 and 1884 there was appropriated, in seven items, the sum of $123,000 for the improvement of this river.

CHAPTER XVII.

CONCLUSIONS AND DEDUCTIONS.

The Supreme Court declares (11 Wallace, 411) that—

1. A river is a navigable water of the United States, when it forms by itself, or by its connection with other waters, a continuous highway over which commerce is or may be carried on with other States or foreign countries in the customary modes in which such commerce is conducted by water.

2. If a river is not of itself a highway for commerce with other States or foreign countries, or does not form such highway by its connection with other waters, and is only navigable between different places within the State, then it is not a navigable water of the United States, but only a navigable water of the State.

In 20 Wallace, 430, it was further decided with reference to the Fox and Wisconsin Rivers, that—

3. Commerce is conducted on the water, even at the present day, through other instrumentalities than boats propelled by steam or wind. And the true test of the navigability of a stream does not depend on the mode by which commerce is, or may be, conducted, nor the difficulties attending navigation. If this were so, the public would be deprived of the use of many of the large rivers of the country over which rafts of great value are constantly taken to market.

It would be a narrow rule to hold that in this country, unless a river was capable of being navigated by steam or sail vessels, it could not be treated as a public highway. The capability of use by the public for purposes of transportation and commerce affords the true criterion of the navigability of a river, rather than the extent and manner of that use.

These, and many similar decisions, which could be quoted, show that breadth of view may be expected from the judicial tribunals, rather than narrow and limiting bounds, when this great subject comes up for decision in any of its details; and such has been the uniform record. So, also, in Congress, we have found that the aim has been a universality of attention, and an intention to promote the welfare of all.

The assertion could, I think, be substantiated by quotations from judicial decisions that it would be right, and just, and proper, in Congress, if it so chose, to make any appropriations that it saw fit, for any navigable stream, or part of it, however small and trifling, without any hope of a return therefrom. But assuming that Congress has always attempted to limit these appropriations to those cases where it was possible that adequate results might be secured, let us now give a summary of our investigations, with this end in view, and ascertain, if possible, wherein the commercial results have been such as to authorize the statement that the work was expedient. To give the foundations for this summary has already been stated in the first chapter to be the chief aim of this essay.

The following general assertions are believed to have been proved by the history of our rivers :

1st. Rivers are not now, whatever they might have been in the past, great through lines of trade. And not only that, it is evident that in all cases the amount of such through business is decreasing, rather than increasing. The records of the Ohio, the Missouri, and the Upper Mississippi, all show that. The large amounts of coal shipped from Pittsburgh down the Ohio seem, but only seem, to be an exception to this rule. It follows as a corollary of the statement that the size and length of a stream, and its connection with other navigable streams, are not the measure of its importance, commercially. Nor, as a further corollary, does the necessity of making appropriations increase with the size of a stream, commercially, if the engineering features do not call for increase.

The relations between Pittsburg and the Upper Missouri, by boat, were formerly important; they are now gone. The long lines of Ohio and Mississippi and Missouri Rivers are not now used as the lines of freight communication. The most valuable freight for its weight, cotton, does not travel up stream for any distance; and the lines of packets which formerly connected New Orleans, Vicksburg and Memphis with Cincinnati, Louisville, Pittsburgh and Wheeling, carrying cotton for eastern shipment by rail, do not now exist. Nor do we find the same boats carrying wheat, corn and flour down the rivers to New Orleans and the ocean, as the easiest and cheapest way to foreign markets. Whatever may be the cause of this state of affairs, it exists. The jetties at the mouth of the Mississippi have not drawn large ships or a numerous fleet to New Orleans, and as much (no more) cotton is now shipped there as before the war. The determined effort

made to induce wheat and flour to go via the Mississippi in 1880 to
1883 did not succeed, and the lines of barges then in use have not had
any great effect upon the course of trade. Moreover the very wheat
and flour which does go south goes by rail, rather than by river.
The following tables have been compiled, and have given me this
conclusion:

STATISTICS OF THE COMMERCE OF NEW ORLEANS.
CLEARANCES AND EXPORTS.

Year.	STEAM VESSELS.		SAIL VESSELS.		Value of Exports.	Percentage to total Exports of Merch'dise from U. S.
	Number.	Tonnage.	Number.	Tonnage.		
1871...	1,201	1,231,161	1,141	566,416	$95,247,791	22
1872...	704	739,850	1,000	457,996	90,802,849	21
1873...	444	473,965	811	517,833	104,615,092	21
1874...	477	485,499	843	453,056	97,086,021	17
1875...	433	459,546	658	335,429	71,062,072	14
1876...	393	446,265	841	439,224	84,228,170	16
1879...	406	591,351	680	368.227	63,795,557	9
1880...	575	810,915	722	397,456	94,152,000	11
1882...	524	834,308	445	231,541	71,021,000	10
1883...	501	565,042	361	188,237	87,298,550	11
1884...	505	544,142	263	123,233	86,122,481	15

Average tonnage of steam vessels 1871–1875......1,050 tons.
"   "   "   "   "   1876.............1,135 "
"   "   "   "   "   1879–1880......1,430 "
"   "   "   "   "   1882.............1,592 "
"   "   "   "   "   1883–1884......1,103 "

The number of clearances from New Orleans in 1860 was 2,185,
with a total tonnage of 1,219,228.

From a study of the value of the cotton crop in the United States,
I have deduced the following percentages of the same, as shipped from
New Orleans: In 1871, 25 per cent; in 1873, 35 per cent; in 1876,
30 per cent; in 1883, 25 per cent; and in 1884, 19 per cent.

SHIPMENTS OF WHEAT AND FLOUR SOUTHWARD FROM
SAINT LOUIS—1873 TO 1883, INCLUSIVE.

YEAR.	WHEAT.		FLOUR.	
	By rail and Miss'pi River. bushels.	Direct to Europe by New Orleans. bushels.	By boat. barrels.	By rail. barrels.
1873			1,266,045	323,078
1874			1,186,698	533,077
1875			797,039	643,641
1876			768,304	436,825
1877			565,475	483,812
1878			565,744	451,913
1879				
1880			644,593	700,849
1881	893,254	4,197,981	632,038	871,386
1882	368,574	5,637,391	726,513	934,968
1883	617,129	1,276,673	553,804	1,080,422

EXPORTS OF WHEAT IN BUSHELS AT CERTAIN
SEA PORTS DURING 1881-3.

PORTS.	1881.	1882.	1883.
New York	41,788,767	37,620,103	21,712,653
Baltimore	19,676,640	17,564,407	15,375,093
Philadelphia	9,008,869	5,838,622	4,416,872
San Francisco	22,306,537	37,132,575	24,386,327
New Orleans	4,420,614	6,100,233	2,222,442
All American ports	96,872,118	113,880,330	75.086,268

Report No. 1571, Senate, 48th Congress, 2d session.

The large increase in coal shipped down the Ohio River from
Pittsburgh, part of which goes to New Orleans, is the principal item
of long distance freight. Although of immense bulk, yet, in value,
this freight is not great. The year of the greatest output, 1882, the
total value, at six cents a bushel, would make it about one per cent.
of the cotton crop for that year, and about six per cent., or a little
more, of the exports from New Orleans, that year.

Now, if rivers themselves are not great through lines, at the present
time, it is not wise to insist that they must be so considered, and,
efforts to make them such, will fail.

2d. All the navigable waters of the United States, whether large
or small, long or short, are valuable lines of communication, in some

way. They either afford the only easy route through some difficult country, mountainous, or swampy; or they give to those who can reach them, the alternative of selecting, by means of cheap transportation, such markets for their products, or sources for their supplies, as may be most advantageous. It may be of as much service to those who live on the Mississippi, to select one of many competing towns, which can be reached by river at short distances, as it is absolutely necessary for the poor citizen, who lives in the rough and mountainous regions of West Virginia, or Tennessee, to use the river in poling his slender stock of supplies against the swift current, in a small push boat.

It is as desirable to the latter to have his small stream freed from obstacles, which his own efforts can not remove, as it is important to the inhabitants of the prairie states, who may use small or great streams, that their navigation be not stopped by low bridges and narrow spans.

In looking over the long lists of rivers, with miles, and miles, of navigable water, upon which float all kinds of products, it is difficult to grasp the idea of the vast numbers of persons and interests affected by their full and free use, if not dependent upon them. It is difficult to analyze, and difficult to sum up, a subject so large and so varied. And, since history, and instincts, the constitution and practice, all combine to select the United States as the guardian and trustee of these interests, it seems to narrow the question to one of expediency, and judgment.

And in the exercise of these fiduciary powers, it must be admitted that it were best to aid a community, sparse in numbers, and poor in resources, and in the surroundings of nature, rather than to give to those who have already received. That it were better to develop the latent wealth of the land, and bring the waste places, and the non-productive regions, to a condition of prosperity and independence, rather than trust to an indefinite hope, has already been shown to be a wise practice in the legislation of our country. In every way has the Nation wisely and generously helped every infant industry in the past, helped to build, or constructed lines of communication, and thus brought the broad lands of the West within reach, filling them with the homes of peace. Surely the time has not yet come to abandon this system.

3d. It is well to note what commodities are found as prominent articles of freight upon the rivers. In all cases where bulky freight can reach a water course, and float down it with the current to a market, it is apt to do so. The more valuable products are apt to follow the railroads, and freight rarely goes up stream, even for short distances. The coal from the Monongahela and the Kanawha goes to Cincinnati and Louisville, and large quantities even to New Orleans, floating in

12

large masses. The steamer J. B. Williams towed thirty-seven pieces in 1883, carrying 712,000 bushels of coal, or a total of 24,480 tons, at once. Gas coal from Pittsburgh will ascend the Mississippi River to Saint Louis. It can not be a visionary idea that, in the future, coal can be taken down the Ohio and Mississippi, and out into the Gulf of Mexico, to replace English coal which now fills coaling stations there, and is brought such a vast distance. Possibly not only coaling stations in the Gulf of Mexico, but on the Atlantic, at no very distant date, may be supplied in the same way.

The coal which lies around the headwaters of the Kentucky and Tennessee streams, and now occasionally comes down on the rivers, may in time be a larger item of the commerce of those streams.

The iron ores, and limestone, and pig iron, already noted as among the freight articles of the Tennessee rivers, may follow the same law. The timber traffic has been very extensively developed on the Upper Mississippi, and an examination of the records given in that chapter will be instructive. It is also to be noted that the importance of the small streams which bring down logs and small rafts is in some instances phenomenal. Such rivers as the Chippewa, Saint Croix and Saginaw are seldom found, and yet in our vast country where may they not be equaled? As business interests oscillate from one to another section, we find such surprising developments that the prophet rarely appears —until after the fact.

It may be noticed that in the records of the southern streams the timber interests are yet in their infancy, and rafts and manufactured timber are rarely noted, except upon a few streams.

For purposes of comparison in these streams, statistics of cotton are generally given. It will be seen by examination that while cotton, in a greater or less amount, is noted upon them, yet the steamers do not develop a larger business, as facilities increase, and the crop is generally moved by the railroads, to which the steamers bring the cotton, over short distances. Cotton is too valuable to be carried long distances by steamer when a railroad can be reached, and yet the sudden and important development of the cotton business upon streams, as soon as they are made navigable, as on the Neuse, Contentnea Creek, Chattahoochee, Noxubee, and others, shows that the river has superseded some more expensive line by which this staple had been carried before.

In no instance, and upon no river, do wheat, corn and flour, in large quantities, go long distances. In general, it is found that agricultural products of all kinds follow the law that is shown in regard to cotton. They may be carried short distances, in varying quantities, and evidently seeking a market. And no river, that can float a steamer, is noted, that may not develop, in an agricultural country, some trade of this kind. In some instances, as on the Wabash, the very varying quantities, and the very different statistics over different parts, all point out, and illustrate this condition. But, to make navigation successful, there must be a fair profit to the carrier, and, after a trial, if such does not follow, then the statistics show it.

Passengers do not now follow the rivers, except in small numbers, for short distances, and there is hardly anywhere shown a well developed summer passenger business. But there are great numbers of short stretches of well populated rivers, where small steamboats ply, connecting the people with the towns, and carrying stores and passengers in large quantities. Such are the Monongahela, Kanawha, parts of the Ohio, parts of the upper Mississippi, and the Bayous of Louisiana; and in these cases, the rivers are especially valuable, and convenient.

The reports are, however, deficient in some cases as to the exact figures of the commerce connected with rivers. It would be better if the annual reports would contain a brief summary, showing for each year, or each part of it, when commerce varies much, the number of steamboats plying on the river, the number of trips made, the tonnage of the steamboats, and of the freight carried. Statistics of rafts, and flatboats, and barges, also of minor boats, should be kept. The expense connected with collecting these statistics should be made a part of the appropriation, and so stated in the bill. But, as seen, these statistics varying very much, several years should be considered together to get a clear view of the changes.

ENGINEERING CONSIDERATIONS.

It is apparent from the known conditions of the rivers flowing west from the Appalachian chain that a general similarity of conditions exists throughout, and that in all cases where navigation exists, there is a limit (generally somewhere near 18″ to the mile) where the slope becomes too great to allow of a navigable depth throughout the entire season. Below that upper limit comes a succession of pools, permitting a very small draft throughout the year, when the channel is entirely contracted to its narrowest possible limits. In extremely low seasons, even this stretch becomes deficient in water, and all navigation ceases. To secure a reliable depth of as much as three feet throughout the year, by contraction methods alone, requires rivers with more water and less slope than can be found upon these rivers, until the lower Ohio is reached. But the problem of their improvement is eminently satisfactory in regard to the permanence of the work done.

The stone wing dams and hard gravel, or rock bottoms, and the freedom from sediment, allow tremendous floods to pass over these works annually, without effect upon the low water channel. Therefore what is done is fairly permanent, and, since it is comparatively inexpensive, satisfactory.

Above the belt of permissible contraction works, when the slope becomes too great, or the amount of flow too small, slack water alone will give permanent navigation; and we find this eminently satisfactory system at work on many streams. The demarcation between the two systems is plain and plainly marked in the case of each river, and no doubt has existed when to apply either, upon the Ohio River and its tributaries. But when, in going down the stream, other conditions are met, and great breadth, masses of moving sand, greater expense, are additional elements, an answer to the problem is not so apparent. Experimental work has been done in the direction of contraction, and in the experimental movable dam near Pittsburgh.

The record has shown that in all cases of canal work, or locks and dams, executed by the officers of the Corps of Engineers, in the past, a success has been secured. In no case have the works been inferior to those done elsewhere, or by others. In no case have such works been recommended unless the necessity has justified the action, and in no case has the money been supplied with the promptness which economy, and the best interests of navigation, warrant; except, perhaps, upon the Kanawha River. The great works at the Falls of the Ohio, the Des Moines Rapids, the Sault Sainte Marie, the Illinois River, the Cascades of the Columbia Canal, have all suffered, for years, from lack of funds, and have never been apace with the needs of navigation. The destruction of coffer dams, the washing away of uncompleted canal banks, the filling up of excavation, waiting for masonry, when owing to suspension of work through lack of means, all add greatly to the estimate of cost of a work, and make the engineering features seem less creditable than they are. A canal, and generally also, a lock and dam, are definite items, the cost of which can be reasonably approximated, and it would be much more economical to make an appropriation in one bill, equal to an entire cost of such work, and then postpone, if necessary, other similar works until a similar entire sum can be given at once, than to divide annually insufficient sums among them. It is greatly to be hoped, for engineering purposes, that, (for example), the sums needed for the completion of the Cascades Canal; or the enlargement of the Sault Sainte Marie Canal; or that of the Falls of the Ohio, could be given in one sum; and better the whole for one at once, than a division among the three.

In leaving the Ohio river, the next step brings us to the upper Mississippi River, where an abundance of water for a low water channel is assured, but the moving sand and shifting channel give instability to a natural condition of the water way.

While the problem here is, in my opinion, still unsettled, because not as yet carried to completion over a sufficient distance, or during a term of years long enough to give every variety of experience, yet the work in progress is such that must be done anyhow, and to make it a success will in any event require only to be supplemented by further work, which possibly the nature of the commerce concerned may not require. That the reservoir system will, if carried out in full, give all the relief that is needed in the upper river, and do all that its advocates claim, can not be doubted. But that the lower portion of the Upper Mississippi will, under contraction methods now in use, have a permanent and deep channel, may well be doubted, as will be discussed later.

In reaching the lower Mississippi and the Missouri, where the great problem of their improvement has been so much discussed, and upon which construction has so far advanced, it is my wish not to criticize, or to seem to do so. But I believe the importance of a clear understanding of the whole will justify me in presenting some views and deductions which the present condition of the question seems to warrant, and which yet could not safely have been predicated before construction.

In reviewing the entire system of rivers we find certain characteristics common to all. Every river is formed of a system of pools of deep water separated by shallower water and bars. In low water these pools become marked, and the shoals and swift water separating the deep and quiet stretches, are essential to the stream. There is no exception to the rule, and the building up of the bars upon the Mississippi and Missouri at higher water, to preserve the status, even at those stages has already been noted. Captain Suter, in describing in 1875 (part 2, p. 506,) the formation and rate of travel of the sand waves down the lower Mississippi, and other features of the river bed says:

The persistency of the principal bars is remarkable; indeed it is probable that many of them are only disturbed by the exceptionally great floods which occur at long intervals, and generally cause great changes throughout the whole length of the river. This persistency can only be accounted for by the sifting agency of the river, which, sweeping away the lighter portions leaves the heavier behind, thus year by year adding to the solidity of the bars. The fact that the bars remain persistent does not conflict with the general motion of the sand waves, as the latter, formed by looser material, constantly coming into the river from caving banks and tributary streams, bury the permanent bars under this moving flood of sand, which on a falling river begins to cut out down to the more solid material. The first motion of the sand waves, caused by a rise, fills up all inequalities in the bed of the river behind the main bars, and even the beds of the pools are greatly raised.

It is always found that during every low water season a limited number of bars show considerably less depth of water than the others, and hence may form the gauge to which navigation must adapt itself. The numerous and accurate surveys made under the Mississippi River Commission have confirmed these views, and show that the Mississippi River is no exception to the general rule. Now, if this be a general characteristic of all rivers at all times, it should be recognized as an essential in all plans of improvement. In 1875 Captain Suter recognized this important fact and said, (l. c. p. 508): All streams are liable to great floods, which carry off in a very short time a large percentage of the annual discharge. During the remainder of the season there are but small additions to the volume of water, and were it not for this formation of the streams they would go dry like ordinary torrents; the shoals hold the water back, and it is stored up in the deep pools to be drawn out gradually as the season advances. On the large streams these pools always contain enough water to maintain fair navigation throughout the dry season, provided it be used with discretion; but any attempt to tap them too soon or too lavishly will, by prematurely draining them, increase rather than remedy the evils complained of.

When an improvement is projected the bars must be looked upon as dams and treated accordingly, it being borne in mind that it is better not to increase the actual flow over the dam but rather to decrease its width and increase its depth, the flow of water remaining the same. No matter how low the Mississippi may be, there is always sufficient water passing to form and maintain a channel of the width and depth deemed necessary for the wants of navigation, if concentrated in a channel of suitable width and depth. All channels through the bars, whether natural or artificial, will be filled up when the sand waves begin to move. The tardiness with which they form naturally is the great impediment to navigation, and the object of any works of improvement must be to hasten and direct this process so that the channel will always be found in the same location, will be formed promptly, and will furnish the depth deemed necessary for navigation. This great principle is not recognized as such in the plans of improvement now being executed upon the Mississippi River. That contraction works alone will not secure the channel as desired is shown by these facts.

[1]One remarkable fact was noted on the 25th of September, 1879, on the Mississippi River, between Saint Louis and Cairo. It was found that the low water of the river was passing through a space 743 feet wide. A still less width was noted but not measured. In each case the channel was bounded on one side by an ordinary sand bar. [2]A second is, that the well regulated river near Carrollton, Louisiana, shows, that under the most favorable circumstances of gentle slope, permanent width, small flood height, and abundance of

---

1. 1880, II, p. 1367.   2. Rep. Miss. C. 1882, p. 111.

water, it does not follow that a channel is permanent in position. But if the channel be not permanent in position, there will be times when it will be deficient in depth; and it be too narrow, and too deep, the pools above will be unduly drained.

The theory in question, is such, that one exception to the rule necessarily vitiates it, and if nature has supplied that, we need not use other argument.

Works of contraction have not proceeded, as yet, to a distance long enough upon the Mississippi River to justify any other assertion than, that the permanence of the bed, within the desired limits, has been secured; the position and dimensions of the channel are yet uncertain; and the preservation of the pools, or, in other words, *the control of the sill of the channel, at the bars,* must be secured, also; and in this way the level of the surface of the low water pools.

But the improvement of the Mississippi, and of the Missouri, depends, also, upon the permanence of the beds. The wonderful bodily movements of these rivers must be first checked before the channel can be defined and secured; and the record of change showed in these pages, of erosion, of cut-offs, and of the destruction of attempts to control this wasteful power, must faintly show how vast the danger, and difficult, and expensive the work. To secure the permanence of the bed of the Mississippi is, then, a separate and greater problem than to secure and maintain a channel.

The works done upon the river are necessary, from this point of view, but, if successful, should not, in my opinion, also be expected to solve the entire problem. It may be well to call attention, here, to the fact, that, while in some cases revetment works have not been successful, it must be admitted that the results obtained from pile work (when free from the attacks of ice), have been very remarkable; and that the control, and deflection of the main channel of the river, in the two great lower reaches, have been unique engineering feats, and, as such, should be recognized.

The problem of the disposition of the sediment of these rivers is a vaster problem than either of the preceding. It is difficult to see how the output of the Missouri can be prevented without the revetment of the banks of that river and of its tributaries, and it is still more difficult to see how the sediment can be kept out of the lower river without holding and maintaining its banks.

When once in the river, nature has disposed of this sediment in but two ways. In time of flood the overflow has deposited on the banks a certain portion, which, building up those banks, finally raised the river to such a height that the head of water secured enabled it to break through its bonds and adopt a new line elsewhere, and resume the same process. The only other disposition made has been at the river mouth, in building up and extending the delta, and the alluvial region, formerly the region of outflow. If the one exit be closed, the other must have additional material brought to it; unless he sources themselves of the sediment be closed also. Now, with an extension of the delta, comes necessarily a diminution of slope, which is to be corrected in time of flood only by shortening the river, as cut-offs have done, or by increased flood heights as seem to be, in the future, if recent experience of the Missouri and Ohio be repeated; either alternative being undesirable. In any event the whole combination of circumstances should be considered, together with all the facts and necessities, and all the gain or loss, and then the cost calculated.

The vastness of the undertaking, the complexities of the situation, the many requirements, make it a whole which should be considered as such; and while there has been nothing to indicate that it cannot be accomplished, yet if undertaken there is every reason to fear excessive loss and disappointment, unless the magnitude of the operations be duly appreciated, and supported with full means and encouragement.

The statistics of commerce upon the lower Mississippi River are wholly wanting, and if the relation between the results and the costs are to be as rigidly required in this as in other cases, it is but just that it be made a part of the duty of the funds for the river to obtain these figures for different points upon it, as are shown in the reports from the Upper Mississippi. But if the entire scheme of general improvement as indicated above be not intended, it would be wiser to omit for the present any effort to improve the lower Missouri River or the lower Mississippi until such time as the finances of the nation, the needs of commerce, and the reclamation and preservation of the vast areas of submersible lands upon these rivers be in a condition to justify an expenditure equal to the magnitude of the scheme

Under any event there remains only the one way to consider these streams, in reference to navigation; and that is that they are in their several parts so many broken and individual lines of communication, useful only to that limited extent in commerce, just as would be smaller streams similarly located. They are not now at the present great through lines of intercourse, nor are they likely to be such in the near future. As local lines they probably furnish all the facilities for navigation that are needed, and are valuable to that extent. The condition and attention received prior to 1878 are the best measures of the present necessities.

CHAPTER XVIII.

THE LATEST REPORTS FROM THE RESERVOIRS; LOWER MISSISSIPPI; AND FROM THE DAVIS ISLAND DAM AND OTHER WORKS.

Since the writing of the preceding chapters, the season of 1885 has passed, and from the officers in charge of several important works I have received information which enables me to give a brief summary of the present condition of these works.

THE RESERVOIRS OF THE MISSISSIPPI.

The three completed reservoirs, (Winnibigoshish, Leech Lake, and Pokegama,) were operated this season in the interest of navigation, by liberating the surplus water that had been impounded, or the water that had been collected over and above what had been steadily liberated to meet the usual needs of the river. Early in August the discharge was increased steadily, and by the end of the month the maximum fixed for the season had been attained; this was kept up until October 18th, when gradual reduction commenced in order to bring it back to the Winter flow, which point was probably reached about November 10th.

The river at St. Paul had been falling steadily for a number of weeks until on the 23d of August it reached a stage of 3′ on the Signal Service gauge there; it having fallen 1′.3 between the 1st and 23d. The zero of this gauge is referred to the oldest known low-water bench at St. Paul; and the gauge has always been followed by Maj. Allen, as the adopted record. The river stopped falling August 23d, and remained stationary until August 31st, when it raised one-tenth; it then gradually rose to 3′.9; which point was reached September 19th. It then gradually got back to 3′.0 by October 28th, where it stood on October 30th. After that time it gradually fell. There are lumbermen's dams upon the Pine and other rivers, which were built and are operated by private corporations at their own will and in total disregard of the public interests, and these have an effect upon the river at times. This Fall the country above St. Paul has been unusually dry, all the tributaries have been low, and generally falling during August, September and October. The Minnesota River discharge dropped lower than ever before measured in summer or fall. Therefore the reservoir discharge had no assistance. For fully eight weeks the contribution to the flow of water past St. Paul, from the reservoirs, was about three-eights of the whole, while *the increment*, from the impounded water, to the low water flow was from thirty to thirty-three per cent. Thus: the average flow past St. Paul being for this period, 8,300 cubic feet, and the average discharge from the reservoirs, 3,000 cubic feet, the ratio is $\frac{3000}{8300}$. The increment over low

13

water flow caused by the reservoirs, is calculated to be 2,200 cubic feet, and therefore the unassisted flow past St. Paul would have been 6,300 cubic feet, and the percentage of gain has been $\frac{2200}{6300}$; at the least.

For ten weeks the reservoir effect at St. Paul was equivalent to increasing the height on the gauge 1'.3 to 1'.5, or in other words the water surface stood at St. Paul 1'.3 to 1'.5 higher than it would have done without the action of the reservoirs.

During the interval August 23d to September 19th, the Quincy gauge showed, first, a rise of five feet in seven days, then a fall of five feet and four-tenths in eleven days, then a rise of two feet and two-tenths; then a steady fall for a month, amounting to five feet; all of which merely represents the general condition of that portion of the river 500 miles below St. Paul.

PLUM POINT REACH—LOWER MISSISSIPPI.

The channel through this reach has presented no obstruction to navigation during the last low water, the depth having never been less than at points above, and the cutting out more prompt and definite; so that while least depths of 9½ feet at Bullerton, and 9 feet at one other point, have occurred, they have continued but a day or two. The characteristic depth at Bullerton Bar has been 12 feet.

The channel has been so tortuous at the foot Island 30, as to make it somewhat difficult to run, especially for boats with tows. The Dikes are mainly in a rather dilapidated condition, especially the earlier ones, owing partly to insufficient initial strength of the first designs, in part to the action of heavy drifts, striking and lodging upon them; somewhat to undermining by the current, and largely to decay. The best and most satisfactory are those closing Osceola and Bullerton chutes.

The revetment works, particularly that of late standard construction present a more favorable aspect. None of that class has sustained any loss, nor been damaged by the action of the river. This applies to both the Memphis and Plum Point experience. At Memphis no other question is involved beyond the mere stability of the works.

LAKE PROVIDENCE REACH.

The contraction works generally, as far as completed, have accomplished the object of their construction. Three large chutes have been closed, many hundreds of acres of bank built up by sedimentary deposit, and the consequent contraction of the water way has improved the shoal crossings very materially.

The least depth found during the past low water season, on any crossing directly influenced by the works of contraction was 14 feet, the Lake Providence gauge reading 7 feet.

Before the works were built, 5, 6 and 7 feet were the ruling depths at the same gauge reading.

The revetment in Pilcher's Point, or Louisiana Bend, has held the general line of the bank against caving.

Soon after a new revetment is put down, weak places, whose positions can not be anticipated, always develope. This was the case at Pilcher's. No funds were available for immediate repairs, consequently, the revetment is in a very ragged condition; it has accomplished, however, so far, the object of its construction.

Little or no damage has occurred to the contraction and revetment work during the past six months, and the channel throughout the reach has remained in good condition.

The report of the Mississippi river commission, covering its operations from October 1, 1884, to the end of the fiscal year (June 30) is made public.

The total cost of bank revetment between Cairo and Vicksburg up to June 30, 1885, has been $2,240,000, and of work for contracting channel, $2,500,000. A very considerable portion of the sum expended for bank revetment was designed to give protection to certain cities and harbors—Memphis, Vicksburg and others. Some of these required special and prompt treatment. If Delta point had not been held by revetting its banks with mattresses at considerable expense, the city of Vicksburg would long before now have been practically an inland town, entirely cut off from the river. At Memphis great values were also put into jeopardy by a rapidly caving bank, which threatened to carry off a portion of the city. Bank revetment, as offering the only possible means of averting the danger, was successfully applied in this case.

It may be stated that it is not the intention, nor has it been the practice of the commission, to protect the bank by revetment merely because it is caving. Other considerations must govern this question. But where an imminent danger threatens the immediate destruction of interests of great value, as, for example, where a caving bend is about to attack in flank and carry away costly works of improvement, or produce a disastrous cut-off, or where a city's river front is to be maintained, as at Vicksburg, or a portion of a city itself is to be protected against undermining, as at Memphis, then it is believed to be imperative that the local remedy of holding the banks intact by a mattress revetment or other equivalent device should be adopted.

The commission says the money percentage of damage and loss in works constructed under the direction of the commission from the time of its organization to June 30, 1885, between Cairo and New Orleans is about 24 per cent. It is in view of serious misapprehensions touching the original plan and recommendations of the commission which have come to its notice that it has thought proper to make this somewhat extended reference to the place which the work of bank protection had in that plan, and to the amount, cost and efficiency of the work of that kind which has been done.

The cost of improving the Mississippi river, the commission says, in the manner and to the extent contemplated, will doubtless considerably exceed the estimate formerly submitted by the commission. For this there are two reasons; first, that it has been found necessary to make use of stronger and firmer, and therefore more expensive methods of constructions than those upon which from the want of experimental knowledge that estimate was based; and, second, that the percentage of loss from floods has exceeded what was formerly thought to be a fair allowance for this contingency. Much of their loss, however, would have been spared had the stronger methods of construction been resorted to at an earlier day, and future loss from this cause may, therefore, in some measure, be avoided.

There must be no just ground for apprehension that the ultimate cost of this improvement will be inordinately great, or will exceed what the government will be fully justified in expending upon a great national work, in the success of which so large and so varied interests are involved. In order, however, that the increased depths already secured upon two reaches of bad navigation may be utilized and made of some practical value, the improvement should be extended up stream and down. Indeed, it cannot be said that the navigation has received any practical benefit whatever as long as the improvements are restricted to localities hemmed in by bad river both above and below. It might be better were no middle course open, to spread each appropriation judiciously over all the six reaches of bad navigation selected for improvement below Cairo, adding a little each year, if practicable, to the available depths of the worst bars, then to confine the work to Plum Point and Lake Providence reaches as heretofore, even if the low river navigation of these two localities should be rapidly deepened to twenty feet and the feasibility of the commissioners' plan thereby be fully demonstrated. The objective point is the improvement of the river, and not the vindication of the agents of the work except as means to an end.

DAVIS ISLAND DAM.

•   This work was completed October 7th.

Of the many varieties of movable dams, mostly of French invention, that known as the Chanoine system, from the name of its inventor, was selected for use at Davis Island, since it combines the elements of economy in construction, durability of parts, and facility of operation best suited to the navigation of the Ohio river, the particular dam taken as a model being that at Port a l'Anglais on the Seine, a few miles above Paris. Many modifications in the details of construction have, however, been necessitated in order to provide for conditions peculiar to the navigation of the Ohio, and especially to provide for the passage of coal tows, a species of navigation unknown in France.

The pass and the three weirs are all alike in construction, the only difference being that the sills of the weirs are one foot, two feet and three feet respectively higher than that of the pass, this being necessitated by the formation of the river bed. The sills of both the pass and the weirs are laid a little below the natural bed of the river, so that when the wickets are down their upper surface will be flush with the bed of the stream, and will have more water on them than can be carried over Horsetail, the bar next below the dam.

The elements of the dam consist of the floor, or sill, and the wicket with its attachments, called the "horse" and "prop." The wickets are rectangular in shape, being twelve feet eleven and a half inches long by three feet nine inches wide. They are placed in an inclined position, however, so that their vertical height above the sill is but twelve feet one and a half inches. The wickets are spaced four feet apart between centers, so that there is an opening of three inches between the edges of any two consecutive wickets, this space being necessary to prevent interference in raising and lowering. It will be seen, therefore, that the dam is not intended to be water-tight.

It is calculated, however, that the ordinary discharge of the river will be sufficient to maintain the pool at the height of the wickets, notwithstanding the waste through these openings. During the lowest stages, however, when the discharge is insufficient to maintain the pool otherwise, boltens, or "needles," will be placed over as many of the spaces as may be necessary to hold the pool up to its normal depth.

In case of a slight freshet in the river, the water will not be permitted to overflow the wickets, but the increased discharge will be allowed to pass through openings made by letting down one or more wickets on the weirs or by placing them "on the swing." In this manner the pool above the dam will always be maintained at a uniform depth, notwithstanding the varying discharge of the river. The entire dam will be lowered as soon as the natural discharge is sufficient to permit coal fleets to pass over the sill of the navigable pass.

OPERATING THE DAM.

The wickets of the navigable pass are to be opened by means of a small iron boat called a "maneuvering boat," the wickets being raised or lowered separately from the upstream side by means of a small windlass attached to the bow of the boat. In raising a wicket, the end nearest the boat is caught by a hook, and the wicket is then drawn up-stream by the windlass and rises in a horizontal position, so as to present the least resistance to the current, until the prop drops into place on the hurter. The wicket is then "on the swing," and it only remains to depress the up-stream end, which is caught by the current and forced into place. To lower a wicket the top is pulled up-stream from the boat, this movement releasing the prop from its hold on the hurter, when the wicket falls at once to its place on a level with the sill of the dam.

The wickets on the weirs are similarly operated except that on this portion of the dam the wickets are handled from a "service bridge" built immediately above, and parallel with, the movable portion of the dam. This bridge is also a "movable" affair, so that, after having served to lower the wickets, it may itself be lowered; the supporting trestles of the bridge are hinged to the floor of the dam, and when lowered they lie overlapping each other on the bottom of the river, their highest parts being on a level with the lowered wickets, and all being protected by the up-stream wall of the "trestle chamber."

THE LOCK.

The lock is built on the right bank of the river, and in area is the largest in the world, its clear interior dimensions being 600 feet by 110 feet, or sufficient to admit a tow-boat with ten barges of coal, and to pass all at a single lockage without breaking up the tow.

The most striking feature of this lock, if we except its unusual size, is in the construction of its gates. The ordinary lock-gates, which consist of two leaves, and are swung into position from the walls of the lock, are in this case replaced by a single sliding or rolling gate, resting upon an iron track, and supported by a succession of iron trucks. The upper and lower gates are similar in construction, and the lock is opened by sliding these gates lengthwise into recesses on the land side of the lock.

The lock is filled by means of fourteen culverts, each four and a half feet in diameter, of which seven are in the river wall, and are supplied directly from the pool above the dam; the other seven, being in the land wall, are supplied from a conduit behind the wall, and are fed from openings in the upper gate recess. The discharge is effected through fourteen valves, three feet two inches in diameter, placed in the bottom of the lower gate, and through seven additional openings, four feet six inches in diameter, in the lower gate recess.

All the machinery of the lock, including the gates and the valves of the filling and emptying culverts, is operated by means of hydraulic jacks and small turbines working under a head of forty feet of water supplied from two large storage-tanks on the bank. The water is pumped into these tanks from the river by a large turbine in the river wall of the lock, working under the natural head of the pool. When there is no head the tanks are filled by a steam pump, but the regular supply is from the turbine.

Hydraulic jacks are used for opening and closing the valves of the filling culverts, and the turbines for operating the gates of the lock. By the aid of this water-power the permanent force required for operating the lock is reduced to two men for the day-watch and two for the night-watch, or four men in all: one man being able, by a quarter turn of a lever, to open or close all the filling valves simultaneously, while by other equally simple motions the ponderous gates are either opened or closed.

When the dam is to be raised or lowered the force will be increased temporarily to six or eight men.

MUSCLE SHOALS CANAL.

On July 1st, 1885, the Elk River Shoals division was excavated completely, one lock ready for the gates, and the second lock had all its stone ready for laying. Of the main canal the excavation is well under way; the locks have twelve iron gates completed, and eight of them in place. Operations are to be continued with the balance on hand. The officer in charge asks $550,000 for the coming year; and the canal can be of no service until completed. The work has been under consideration for 17 years, and under construction for 10 years; the delay arising solely from lack of funds—(see page 51).

SAINT MARY'S FALLS CANAL.

Statistics show that the commerce using this route is increasing so rapidly, that if it continues at the same rate for five years, the present lockage system at the canal will not be sufficient to pass all the vessels. It is now proposed to give the canal a depth of 20 feet, and to replace the old locks, with a new lock with 21 feet on the mitre sills, and a lift of 18 inches. The statistics show that vessels 5,629 with tonnage of 2,981,786; carrying 2,870,728 tons freight, used the canal during the season. Principal freight, coal 691,174 tons; copper 36,829 tons; iron ore, 1,112,828 tons; pig and other iron, 63,083; flour 1,334,802 bbls; and grain 14,130,448 bushels—(see page 13).

CASCADES CANAL, COLUMBIA RIVER OREGON.

The construction during the year consisted in canal walls, excavation, grading, quarrying of stone, and rock blasting in the rapids and along the shore of the Lower Columbia. But as the appropriation for 1884 was only $150,000 and the estimate for completion of the existing project is $1,250,000; of which $750,000 was asked for the coming year, it is evident that the sum of $150,000 available for two years can be of little service, and the canal can not be said to be fairly under construction—(see page 122).

CHAPTER XIX.

FINAL REMARKS.

It has not been my intention to include herein all the streams upon which money has been expended, nor all the points for which appropriations have been made. There are, undoubtedly, streams whose improvement has been attempted without a reasonable hope of success, and without adequate results. Mountain torrents have received money and work, which do not appear to have been made of service thereby; and work has been done at isolated points for securing individual or local benefits, which perhaps are not, in a true sense, improvements of the navigation of the rivers. Also, there are uncompleted projects which may not ever be of value, because not completed. So also, there are some minor streams deficient in water, and in size, which should not be called navigable streams, even in a liberal construction. It would not be wise, nor gracious, in any one to so criticise the action of Congress, as to make it appear that the liberality and generosity displayed by it, were but instances of ignorance, or fraud, or carelessness. Compared with the failures in business of able men, or miscalculations of corporations in their plans, the failures of Congress in the river and harbor system, are few and small, by proportion, and in all cases the intention has been good.

A candid reasoner will also admit that the system of representation is the true one, in this, as in other matters coming up for the action of the general government. In no other way, can the specific needs of a congressional district be so well expressed, as through, and by its representative in the House. But if after due consideration of the interests of the district, the horizon be so limited that the true proportions of objects beyond the near vicinity can not be appreciated; or if in order to secure attention to the needs of the district a just discrimination as to other claims is not exercised, then will the system fall to the ground;—and it should do so.

The history sketched in these pages has not been studied rightly; or I have not presented the facts clearly if these three statements are not facts:

1st. The measure of the commercial importance of a river is not its size; great rivers are not of great commercial value of necessity; nor small rivers of small value.

2d. The necessity of making appropriations for the improvement of rivers is not dependent necessarily upon the existence of a great commerce, or in a wealthy community;—poverty, and necessity go together;—it is better to aid than to be only a competitor;—the small streams in the difficult countries are sometimes the most valuable—relatively.

3d. Conditions change. That an item has been upon the statute once does not require that it should always be there, and occasions arise for aid when a small sum then, will be worth a much larger at another time.

In no other way can the adjustment of the numerous, and varied needs thus represented, be so well done, as by the proper committee of the House, which can make such a decision, as to expediency, as will be forcible and just, and support their views by proper explanations. No delegation of these powers to executive officers, or to commissions of any kind, or sub-committees, can be of value, other than to collect, and present information, upon which intelligent action can be based, and generally the selection of such commissions is resorted to, only to secure postponement of action.

The true combination of judicial and legislative functions exhibited by such committees has always been recognized as of force, and wise; whenever based upon candid, and full presentation of the facts, and exercised in an impartial way.

In reference to the principles of the improvement of the rivers discussed, it cannot be said that they are not understood by those in charge of the work. There is nothing to show that in the history of the past eighteen years, any great mistakes have been made in theory or in practice. Many theories are proposed, many views entertained, and many criticisms made; but in actual practice it cannot be shown that wrong theories are followed in preference to right, or in ignorance of right.

Errors arise due to the necessity of the case which compels the beginning of work, in many instances, with inadequate means, in the hope that the future will not be as the past. More often, however, trouble arises from inadequacy of funds, and the reports are full of the complaints made by those who are executing a large work, requiring an energetic prosecution, and ample funds, to do it to advantage; and yet having an entirely inadequate sum. In business life it is generally thought wise not to begin such a work until at least funds are provided to attain a natural stage where results could be secured, or no loss occur because of lack of means. Yet this is not the law or the custom in many of our public works, and disasters, and delay follow.

If the Secretary of War, the Chief of Engineers, or Engineer Officers, had the authority to recommend and suspend the beginning of work in all cases, until ample funds were assured, there can be no doubt that the record in many points would be different. But although in extreme cases this right has been exercised by the Secretary of War, he is of course very chary of thus putting a practical veto upon the action of Congress; and he has never suggested that such authority would be desirable.

It were a better solution to appropriate at once an adequate amount, or to postpone the work until such requisite sum can with safety be given.

There is also a consideration that may well be brought to mind now. In the short sessions of 1883 and 1885, the bills failed because time did not allow of a full discussion either in the House or in the Senate, but more especially in the latter. It was also unfortunately a fact that the size of the total of the bill led many to think that individual items had been granted in proportion to that total, and that consequently the failure of the bill would not be of the same importance to those items as it would have been had the bill been smaller. But such was not the case and work is now suspended for a year, at a great cost to some permanent works uncompleted, besides many recurring works needing constant funds. Would it not be well to recognize the fact that the short session would interfere with the progress of the bill, and then pass during the long session appropriations to cover TWO YEARS for those works whose continuance is desired, and leave for the short session only such items as can not be anticipated or to meet such emergencies as may have arisen during the year? Engineers would in such event be more apt to arrange their working plans economically, than if it were probable equal additional amounts were to come the next year.

A few words of explanation may be necessary to those who are not familiar with the mode of selection of the members of that corps which has been placed in charge of the improvements of rivers and harbors, conducted by the government.

The Corps of Engineers is formed by successive annual additions from those graduating near the head of the class at the Military Academy. The selection, in accordance with a general rule, of cadets from all parts of the United States, and the subsequent arrangement according to the order of merit, as determined in a four years course of study at West Point, under a permanent system, and by those who are impartial, and unprejudiced, gives, in all probability, a winnowing out, as it were, and a perfect mode of selection, not equaled in the land, from the material at hand. While it is not likely that this course will secure the very best talent in the country, for this, or any other purpose, yet it is certain that no better mode is at present known for selecting a body of men for special purposes, who will be so free from partiality, partisanship, or political favoritism; and who will be so apt to give a strict attention to duty.

And it is by no means evident that the process has not also secured talent amply capable of meeting all the requirements ever made of the Corps, or likely to be made. Of the one hundred and nine officers now in the Corps of Engineers, the record shows that one hundred and three were born in twenty-five different States, and Territories, and six were born abroad.

INDEX.

www.ingramcontent.com/pod-product-compliance
Lightning Source LLC
Chambersburg PA
CBHW021709210326
41599CB00013B/1583